LIFE

Man in Space

An Illustrated History from Sputnik to Columbia

United States
USA

Special thanks to Mike Gentry at NASA, and to Bozena Bannett, Robert Dente, Gina Di Meglio, Anne-Michelle Gallero, Peter Harper, Suzanne Janso, Robert Marasco, Natalie McCrea, Jessica McGrath, Mary Jane Rigoroso, Steven Sandonato, Niki Whelan

Volume 3, No.3 • March 17, 2003

Front Cover Ed White, first American space walk, 1965. NASA

Back Cover Explorer 1, 1958. U.S. AIR FORCE. *Columbia*, 2003. NASA

Half Title Bruce McCandless, first untethered space walk, 1984. NASA

Title Orbiter *Columbia*'s first rollout, Cape Canaveral, 1981. KEN SHERMAN

Right Earthrise from *Apollo 8*'s lunar orbit, 1968. WILLIAM ANDERS/NASA

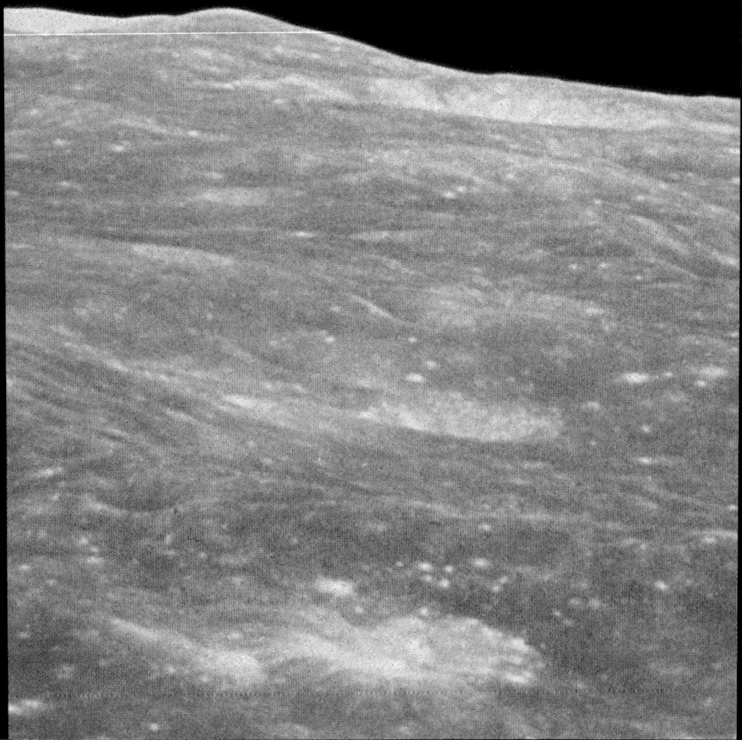

Introduction by Richard B. Stolley

Yesterday, Today ... and Tomorrow?

The fiery track of the doomed *Columbia* recalled, for me at least, the many times in the past that I had watched space rockets disappear in heart-stopping explosions, raining debris onto the earth below. Back then, however, there was one infinitely important difference. No human beings were aboard.

In fact, nearly a half century ago, when the United States entered the space age, the thought of putting men—women weren't yet in the equation—on top of those massive firecrackers seemed appalling.

But we did, and soon. In 1959 the Mercury astronauts were introduced to America, those seven pioneers whose names we quickly memorized, like the roster of a favorite sports team. The first American went into space in 1961—a modest, 15-minute, up-and-down flight—but only eight years after that, our countrymen were walking on the moon. In retrospect, our progress was breathtaking; at the time, difficult and glacial.

Space exploration in America actually began in the 1920s in a Massachusetts backyard where the visionary Robert Goddard startled neighbors by popping off a yard-long rocket to prove that Newton's third law about action and reaction was valid. But our country got serious about the science on an isolated Atlantic beach due east of Orlando, in the cold war '50s. The place was called Cape Canaveral. A few swampy miles south of it lay Patrick Air Force Base and the village of Cocoa Beach. Mosquitoes vastly outnumbered people.

At first the emphasis was on building military weapons. The Air Force wanted an ICBM—intercontinental ballistic missile—that could drop a nuclear warhead onto the Kremlin. The Navy was working on an ICBM that could be fired from a submarine under water. And the Army coveted medium-range missiles for battlefield deployment.

But to demonstrate to the world that we were a peace-loving nation, we also set out to put a benign communications satellite into orbit around the globe. That job, inexplicably, was given to the Navy, which had the least experience in rocketry. The unspoken reason had to do with controlling the bitter interservice rivalry—the Navy got the prize because it felt neglected compared with the more space-savvy Air Force and Army.

By the mid-'50s, these rockets and others were being tested at Cape Canaveral, and a handful of journalists began showing up to cover what we came to realize was the dawn of the space age. As chief of the Atlanta bureau, the closest LIFE outpost, I was among them. Because these were military operations, our scrutiny was not welcomed. In fact, we were considered little better than spies, giving information, aid and comfort to the Soviet Union, which of course was developing its own rocket arsenal in total secrecy. When we tried to report on and photograph a launch, MPs were sent up and down the beach in jeeps to force anyone with a camera to leave. LIFE was a special target.

The photographers and I took to hiding in the palmetto jungle beyond the beach, swatting giant skeeters and baking in the Florida sun through the interminable delays endemic to rocket launches. We had almost no information beyond what we could pry out of hard-drinking Cape Canaveral workers who gathered every night in the bar of the Starlite Motel in Cocoa Beach, and what activity we could observe on the Cape launchpads two or three miles away.

One day we discovered a big house under construction near the beach. No one was at the site. We climbed up the wooden frame to the top of the second floor and had an unobstructed view of the impending launch. As the excited photographer unpacked his tripod, a car drove up and the driver jumped out screaming obscenities at us. Our choice of vantage point was very unwise. The enraged man was the security director at the Cape; this was his house, and he had been tipped off about trespassers by a patriotic neighbor. We barely escaped arrest that day.

What we witnessed so often, launch after launch, was fiery

Robert W. Kelley

Stolley (right) was present at the creation of the U.S. space program, when it was little more than a series of exploding rockets. By 1959, when he posed at Cape Canaveral alongside *Time* magazine's William Shelton, NASA's efforts were up and running, and growing more sophisticated by the day. Satellites and capsules bearing monkeys were being sent into space, and the first manned U.S. flight was in sight.

Bettmann/Corbis

failure. It came in two or three versions. After ignition, the missile would rise a few feet, hesitate and then settle slowly, agonizingly, back onto the pad and vanish in a huge bright orange balloon. Seconds later, a massive boom would confirm what our eyes almost refused to acknowledge.

Sometimes the explosion would come after the missile had cleared the pad, just as our hopes were soaring with it. Other times, the missile would veer off course and an engineer in the heavily fortified underground mission control nearby would punch the "Destruct" button, and fragments of the million-dollar machine would fall into an ocean that had been carefully cleared of boats beforehand.

We few reporters would often gather after a launch, and setting aside competitive instincts for a change, compare notes and trade observations. Such a session saved the hide of one newly arrived reporter who had just called his newspaper with a vivid description of the event. "Wasn't it spectacular when that second stage kicked in?" he burbled. We looked at him. "Pal," someone finally said, "that was not the second stage. That was the damned thing blowing up." He forlornly hurried off to a pay phone.

The press was finally allowed on the Cape when operations switched from military to civilian. The Navy was trying out its new Vanguard rocket meant to carry the first American satellite into orbit. The tests were uniformly, depressingly unsuccessful. One nighttime launch had to be aborted when the Vanguard disastrously headed back toward Orlando. It looked as if the wreckage were tumbling down directly upon the bleachers where we watched, and a few inexperienced reporters ran shrieking into the dark. The only casualties were perhaps a few fish.

No one followed the embarrassing Vanguard difficulties more closely than a stocky, handsome rocket scientist named Wernher von Braun. He was chief of development of the Army Ballistic Missile Agency at Redstone Arsenal in Huntsville, Ala. Widely known as the head of the Nazi program that dumped deadly V-1's and V-2's into England during World War II, von Braun surrendered in 1945 to the American army and was whisked to the United States. There, in Alabama, he eventually surrounded himself with so many wartime colleagues that they sometimes conducted their meetings in German.

In 1952, von Braun had urged that the U.S. build a "manned satellite to curb Russia's military ambitions." Washington finally approved a tiny, unmanned satellite but assigned the task to the Navy. Von Braun wanted to do it himself, was unhappy and often said so.

One of the people he said so to was me. In November 1957, photographer Walter Sanders and I spent a week with von Braun in Huntsville for a LIFE story. The month before, the Soviets had stunned the world—and scared and humiliated the West—by orbiting a 184-pound, basketball-size satellite called Sputnik 1. Americans could hear its "beep, beep" on their radios, and see it streak across the night sky above them. And we hadn't even managed a successful launch of Vanguard yet.

The last morning of my visit to Huntsville, von Braun called me in my motel room. "Have you heard?" he demanded. "The Russians have launched a second satellite. My God, it's as big as a Volkswagen!" He asked if he could come to the motel, and for

NASA

David Hume Kennerly

Opposite: A Juno-11 carrying a 90.5-pound satellite has run wild after launch and is blown up by Cape Canaveral's Range Safety Office. It is representative of the failures witnessed more than once by Stolley in the '50s. His career as a writer and editor has brought him in and out of LIFE through the years; he was managing editor from 1982 to '86. Left: In 1988, Stolley returns to the halcyon days of Project Mercury when he reconnects for a LIFE cover story with John Glenn, who, at age 77, is preparing for his return to space. Above: The space shuttle *Discovery*, with Glenn aboard, lifts off from Canaveral.

the next two hours, the usually diplomatic von Braun paced my room, condemning the U.S. space program in astonishingly blunt language.

Among other complaints was his accusation, absolutely correct, that "obstructive inspections and monitoring" were delaying the development of missiles needed to achieve the cold war balance of terror. At one point he looked at me and almost shouted: "I just wish someone had the authority to tell me, 'All right, we'll leave you alone for two years, but if you fail, we're going to hang you.'" It was an impressive suggestion from someone who had managed to survive under Hitler.

For starters he was asking that he and his team at Redstone be assigned immediately to send up our own satellite. Badly shaken by the Russians' dual achievements, the Defense Department relented. Von Braun used his reliable Jupiter-C missile (which was itself adapted from his V-2) as the launch vehicle, and in January 1958, Sputniks 1 and 2 were chased into orbit by a made-in-the-U.S.A. competitor named Explorer. The space race wasn't lost after all.

Eventually, of course, that race was won—by the United States. And then the Soviet Union was gone altogether, and the other rocket guys were called Russians. Space turned out to be a place where former enemies got along just fine, sharing close quarters in one another's command modules and, later, space stations. In the 1980s, the U.S. replaced its capsules with reusable shuttles, and the dream of hopping off to other planets began to look like reality. The shuttle program recovered from its first tragedy, and appeared strong again. John Glenn certainly had faith in it, as he returned to orbit, for science and maybe just for old time's sake.

Now there's been a second tragedy with the shuttles. The fiery track of the doomed *Columbia* recalled, for me at least, the beginnings of it all, and the many other fantastic images from down through the years—images that fill this book, and bring back the danger and the thrill. They remind us of yesterday, and make us wonder what the future will bring. They remind us of why we went into space, and why we will go again.

Man in Space

50 Years, 100 Landmarks

Blue entries indicate United States space missions.
Red items indicate missions of the former Soviet Union and then, from 1992 forward, Russia.

Here and throughout this book, manned missions and spacecraft are in *italic*. Unmanned missions are in roman, as are the names of space stations, which were unmanned when they were deployed.

The duration of a mission is given when it is deemed noteworthy. Unless otherwise indicated, dates correspond to mission launches.

Redstone rocket: Paul Schutzer

1953

Aug. 20 **Redstone Missile 1** Fired by Army Redstone Arsenal team at Cape Canaveral, Fla.

1957

Oct. 4 **Sputnik 1** First man-made Earth satellite.

Nov. 3 **Sputnik 2** First satellite to collect biological data from orbit; carried dog Laika.

1958

Jan. 31 **Explorer 1** First U.S. Earth satellite.

1959

May 28 **Jupiter** First primates in suborbital space, monkeys Able and Baker.

Oct. 4 **Luna 3** First lunar probe to photograph moon's far side.

1960

Aug. 19 **Sputnik 5** First return of orbited animals from space, dogs Belka and Strelka.

1961

Jan. 31 **Mercury-Redstone 2** First test of Mercury-Redstone flight vehicle with passenger on board, the chimpanzee Ham.

Apr. 12 ***Vostok 1*** First manned space flight; Yuri Gagarin; one orbit; 1 hr., 48 mins.

May 5 ***Mercury-Redstone 3***, a.k.a. ***Freedom 7*** First U.S. manned flight; suborbital; Alan Shepard; 15 mins., 28 secs.

July 21 ***Mercury-Redstone 4***, a.k.a. ***Liberty Bell 7*** Manned suborbital flight; Gus Grissom; 15 mins., 37 secs.

Aug. 6 ***Vostok 2*** Second U.S.S.R. flight; Gherman Titov; 16 orbits; 25 hrs.

1962

Feb. 20 ***Mercury-Atlas 6***, a.k.a. ***Friendship 7*** First U.S. manned orbital flight; John Glenn; three orbits; 4 hrs., 55 mins.

Oct. 3 ***Mercury-Atlas 8*** First telecast from orbit; Wally Schirra; six orbits; 9 hrs., 13 mins., 11 secs.

1963

May 15 ***Mercury-Atlas 9*** First U.S. manned flight to exceed 24 hours; Gordon Cooper; 22.5 orbits; 34 hrs., 20 mins.

June 14, June 16 ***Vostok 5*** and ***Vostok 6*** Dual missions; Valery Bykovsky in *Vostok 5;* Valentina Tereshkova, first woman in space, in *Vostok 6; Vostok 6* came within three miles of *Vostok 5*.

Tereshkova medal: David King collection

1964

Oct. 12 ***Voskhod 1*** First three-man orbital mission; Konstantin Feoktistov, Vladimir Komarov, Boris Yegorov; 16 orbits; 24 hrs., 17 mins.

John Glenn with President Kennedy: Michael Rougier

Nov. 28 **Mariner 4** First successful flyby of Mars.

1965

Mar. 18 ***Voskhod 2*** First space walk: Alexei Leonov performed 10-minute Extra Vehicular Activity, or EVA; 17 orbits; 26 hrs., 2 mins.

Mar. 23 ***Gemini 3*** First U.S. two-man crew; first manned orbital maneuvers; Gus Grissom, John Young.

June 3 ***Gemini 4*** First U.S. space walk: Ed White (EVA, 36 mins.); 62 orbits; 4 days, 1 hr., 56 mins., 12 secs.

Dec. 4, Dec. 15 ***Gemini 7*** and ***Gemini 6*** First rendezvous in space; *Gemini 7:* Frank Borman, Jim Lovell; record 206 orbits; 13 days, 18 hrs., 35 mins; *Gemini 6:* Wally Schirra, Tom Stafford.

1966

Jan. 31 **Luna 9** First soft-landing on moon; returned lunar-surface photos.

May 30 **Surveyor 1** First U.S. soft-landing on moon; sent back 11,240 pictures.

Nov. 11 ***Gemini 12*** Final *Gemini* mission; three EVAs, for record total of 5 hrs., 30 mins.; Buzz Aldrin, Jim Lovell; 63 orbits; 3 days, 22 hrs., 36 mins.

1967

Jan. 27 ***Apollo 1*** Fire inside command module during ground testing resulted in the death of three astronauts: Roger Chaffee, Gus Grissom, Ed White; first fatalities in U.S. spacecraft.

Apr. 23 ***Soyuz 1*** First manned test flight of new *Soyuz* spacecraft; Vladimir Komarov killed in crash after parachute lines of reentry module were fouled.

1968

Dec. 21 ***Apollo 8*** First manned orbit of moon; sent back famous images of earthrise; William Anders, Frank Borman, Jim Lovell; 10 lunar orbits; 6 days, 3 hrs.

1969

Jan. 14, Jan. 15 ***Soyuz 4*** and ***Soyuz 5*** First docking of two manned spacecraft; *Soyuz 4:* Vladimir Shatalov; *Soyuz 5:* Yevgeni Khrunov, Boris Volynov, Alexei Yeliseyev; remained linked for 4.5 hrs.

July 16 ***Apollo 11*** First manned lunar landing; first men to walk on moon, Buzz Aldrin and Neil Armstrong, spent 22 hours on surface with EVA of 2 hrs., 35 mins.; Mike Collins stayed in lunar orbit.

Nov. 14 ***Apollo 12*** Second manned lunar landing; Alan Bean, Pete Conrad, Richard Gordon; on moon, 31 hrs., 31 mins.; EVA, 7 hrs., 45 mins.

Apollo 11 (Buzz Aldrin with flag): Neil A. Armstrong/NASA

1970

Apr. 11 ***Apollo 13*** Third lunar landing attempt, aborted owing to oxygen-tank explosion in service module; crew returned safely; Fred Haise, Jim Lovell, Jack Swigert; 5 days, 22 hrs., 54 mins.

Aug. 17 **Venera 7** First successful landing on Venus.

1971

Apr. 19 **Salyut 1** First orbiting space station and laboratory; deorbited Oct. 11.

May 30 **Mariner 9** First successful Mars orbiter; transmitted 7,329 images of surface and moons.

June 6 ***Soyuz 11*** First crew to occupy orbiting Salyut space station (22 days); because of accidental depressurization, all three cosmonauts died on reentry: Georgi Dobrovolskiy, Viktor Patsayev, Vladislav Volkov.

July 26 ***Apollo 15*** Fourth manned lunar landing; first use of manned Lunar Roving Vehicle; Jim Irwin, Dave Scott, Alfred Worden; 67 hours on moon.

1972

Mar. 2 **Pioneer 10** First flyby of Jupiter; first probe to escape solar system; craft is now over 7.6 billion miles away, headed toward Aldebaran in constellation Taurus, which will be reached in 2 million years.

Dec. 7 ***Apollo 17*** Sixth and last *Apollo* lunar landing; collected 243 pounds of lunar samples; Gene Cernan, Ron Evans, Jack Schmitt; record 75 hours on moon; EVA, 22 hrs., 4 mins.

1973

Apr. 5 **Pioneer 11** Launch of Jupiter probe; first successful flyby of Saturn.

Viking launch: NASA

May 14 **Skylab** First U.S. space station placed in Earth orbit; damaged during launch and required repair.

May 25 ***Skylab 1*** First crew to occupy Skylab (28 days, 49 mins.); replaced thermal shield and repaired solar wing; Pete Conrad, Joe Kerwin, Paul Weitz.

Nov. 3 **Mariner 10** First craft to use gravity of one planet (Venus) to reach another (Mercury); first TV pictures of Mercury.

Nov. 16 ***Skylab 3*** Last crew to occupy Skylab; observed comet Kohoutek; longest Skylab mission (84 days, 1 hr., 16 mins.); Gerald Carr, Edward Gibson, William Pogue.

1974

July 3 ***Soyuz 14*** Crew rendezvoused and docked with Salyut 3 space station for 14 days; Yuri Artyukhin, Pavel Popovich.

1975

Jan. 11 ***Soyuz 17*** Crew docked with and occupied Salyut 4 station for 29.5 days; Georgi Grechko, Alexei Gubarev.

June 8 **Venera 9** Venus probe soft-landed and returned first TV pictures from surface.

July 15 ***Apollo-Soyuz*** First cooperative international mission; docked in Earth orbit for two days; Vance Brand, Deke Slayton, Tom Stafford (U.S.); Valeriy Kubasov, Alexei Leonov (U.S.S.R.).

Aug. 20, Sept. 9 **Viking 1** and **Viking 2** Mars orbiters and landers; first two successful Mars landings; Viking landers returned over 1,400 images from Martian surface.

1977

Aug. 20 **Voyager 2** and Sept. 5 **Voyager 1** Flyby probes launched to gas-giant planets; Voyager 2 returned pictures and data on Jupiter, Saturn, Uranus, Neptune; Voyager 1 studied Jupiter, Saturn and nine of their moons.

1978

June 15 ***Soyuz 29*** Docked with Salyut 6; performed two-hour EVA to replace equipment and retrieve package exposed to space for 10 months; Alexandr Ivanchenkov, Vladimir Kovalyonok; 139 days, 14 hrs., 47 mins.

1980

Apr. 9 ***Soyuz 35*** Eighth crew to occupy Salyut 6; deployed radio telescope; Leonid Popov, Valeri Ryumin; 184 days, 20 hrs., 12 mins.

Nov. 27 ***Soyuz T-3*** First *Soyuz* in nine years to carry three crewmembers: Leonid Kizim, Oleg Makarov, Gennadi Strekalov.

1981

Apr. 12 ***STS-1*** First orbital test flight of space shuttle *Columbia;* Bob Crippen, John Young; 2 days, 6 hrs., 20 mins., 53 secs.

Oct. 30 **Venera 13** Launch of Venus orbiter and lander; first color pictures.

1982

Mar. 22 ***STS-3*** Third orbital test flight of space shuttle *Columbia;* first manipulation of payload in cargo bay; Gordon Fullerton, Jack Lousma; 8 days, 4 mins., 46 secs.

Aug. 19 ***Soyuz T-7*** First coed crew (two male, one female); docked with Salyut 7; Leonid Popov, Svetlana Savitskaya, Alexandr Serebrov.

Nov. 11 ***STS-5*** Fifth flight of space shuttle *Columbia;* first non–test flight launched two commercial satellites; Joseph Allen, Vance Brand, William Lenoir, Robert Overmyer.

1983

Apr. 4 ***STS-6*** Sixth shuttle flight; first flight of *Challenger;* primary payload, TDRS-1 satellite; Story Musgrave and Donald Peterson made first shuttle space walk (4 hrs., 17 mins.).

June 18 ***STS-7*** Second flight of *Challenger;* Sally Ride became first American woman in space; two communications satellites deployed.

Aug. 30 ***STS-8*** Third flight of *Challenger;* first flight of African American astronaut, Guy Bluford Jr.; multipurpose satellite deployed.

Nov. 28 ***STS-9*** Ninth shuttle flight; *Columbia* orbiter; first orbital Spacelab (73 investigations); first non-U.S. crewmember, German Ulf Merbold of the European Space Agency; first time six persons launched in single vehicle.

Discovery launch: John W. Young/NASA

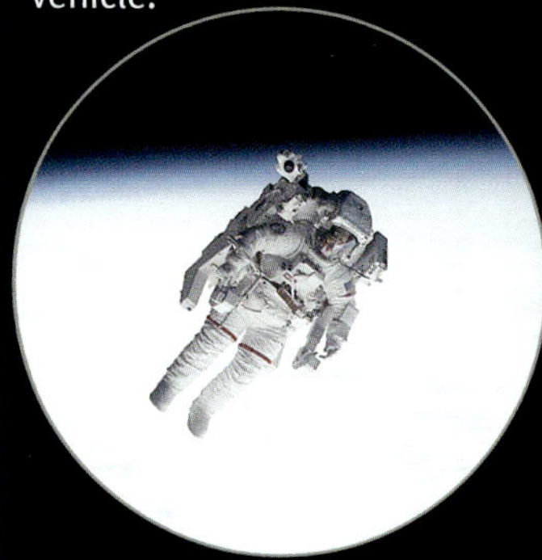

McCandless space walk: NASA

1984

Feb. 3 ***STS-41B*** Orbiter *Challenger;* first untethered space walks, by Bruce McCandless and Bob Stewart.

Feb. 8 ***Soyuz T-10*** Docked with Salyut 7 station; crew set duration record of 237 days and returned in *Soyuz T-11;* Oleg Atkov, Leonid Kizim, Vladimir Solovyov.

July 17 ***Soyuz T-12*** Svetlana Savitskaya became first woman to walk in space.

Aug. 30 ***STS-41D*** First flight of *Discovery* orbiter; 6 days, 56 mins., 4 secs.

Oct. 5 ***STS-41G*** First seven-person crew; first flight to include two women (Sally Ride, Kathy Sullivan); first American woman to walk in space (Sullivan).

1985

Apr. 12 ***STS-51D*** Orbiter *Discovery* carried two communications satellites; first flight of U.S. elected official (Senator Jake Garn).

July 2 **Giotto** European Space Agency flyby mission to comets Halley and Grigg-Skjellerup.

Oct. 3 ***STS-51J*** First flight of *Atlantis,* fourth and final member of original shuttle fleet; 4 days, 1 hr., 44 mins., 38 secs.

1986

Jan. 28 ***STS-51L*** Orbiter *Challenger* exploded 73 seconds into flight; all seven crewmembers died; crew included the first teacher in space, Christa McAuliffe, as well as teammates Gregory Jarvis, Ron McNair, Ellison Onizuka, Judy Resnik, Dick Scobee, Mike Smith.

1987

Dec. 21 ***Soyuz TM4*** Carried crew to Mir space station; returned *Soyuz TM-6* crew to Earth; Musa Manarov and Vladimir Titov completed longest flight to date: 365 days.

1988

Sept. 29 ***STS-26*** Orbiter *Discovery;* marked return of shuttle fleet to operational status after *Challenger* tragedy; 4 days, 1 hr., 11 secs.

1989

May 4 ***STS-30*** Orbiter *Atlantis* launched Magellan, Venus radar mapping spacecraft.

Oct. 18 ***STS-34*** Orbiter *Atlantis* launched Galileo probe to Jupiter.

1990

Apr. 24 ***STS-31*** Orbiter *Discovery* launched Hubble Space Telescope.

1992

Mar. 17 ***Soyuz TM-14*** First piloted mission of the successor space program to the U.S.S.R.'s; Klaus-Dietrich Flade, Alexandr Kaleri, Alexandr Viktorenko flew as Russians, not Soviets; 145 days, 14 hrs., 10 mins.

May 7 **STS-49** First flight of space shuttle *Endeavour, Challenger*'s replacement; 8 days, 21 hrs., 17 mins., 38 secs.

Sept. 12 **STS-47** On *Endeavour,* Mae Jemison became first African American woman in space.

1993

Oct. 18 **STS-58** Orbiter *Columbia* made first two-week shuttle mission; second Spacelab Life Science mission.

Dec. 2 **STS-61** Orbiter *Endeavour* serviced and repaired Hubble Space Telescope; record five EVA sorties.

1995

Mar. 14 **Soyuz TM-21** Carried Mir crew, including Norm Thagard, the first American to be launched in a *Soyuz.*

1996

Mar. 22 **STS-76** Third docking between *Atlantis* and Mir space station; transferred American Shannon Lucid to Mir; Lucid began marathon mission of 188 days.

Nov. 7 **Mars Global Surveyor** Mars orbiter has sent back over 93,000 images.

Nov. 19 **STS-80** Orbiter *Columbia;* longest shuttle flight to date; 17 days, 15 hrs., 53 mins., 18 secs.

Dec. 4 **Mars Pathfinder** Lander and rover would bounce onto surface of Mars on July 4, 1997.

1997

June 25 **Mir-Progress M-34** Unmanned Progress supply module filled with garbage struck aging Mir space station during remote-control docking test, depressurizing part of Mir and resulting in 50 percent power loss for the station; Mir crew, including American Mike Foale, uninjured.

International Space Station: NASA

Sept. 25 **STS-86** Seventh *Atlantis*-Mir docking; flyaround of Mir to help locate space station damage; return of Mike Foale after 145 days in space.

1998

Jan. 22 **STS-89** Orbiter *Endeavour* made eighth Mir-shuttle linkup; Andy Thomas transferred: last American to live aboard Mir.

Oct. 29 **STS-95** Orbiter *Discovery;* return of Senator John Glenn to space to explore similarities between aging process and space-flight experience; 8 days, 21 hrs., 44 mins.

Dec. 4 **STS-88** Orbiter *Endeavour* crew began assembly of International Space Station (ISS).

1999

July 23 **STS-93** Orbiter *Columbia;* Eileen Collins became the first female mission commander; deployed Chandra X-ray Observatory.

2000

Oct. 11 **STS-92** Orbiter *Discovery;* 100th space shuttle flight; continued assembly of ISS.

Oct. 31 **Soyuz TM-31** Expedition One team is first crew aboard International Space Station: Yuri Gidzenko, Sergei Krikalev (Russia); William Shepherd (U.S.); 186 days, 21 hrs., 49 mins.

2001

Mar. 8 **STS-102** Orbiter *Discovery* delivered Expedition Two crew (Susan Helms, Yuri Usachev, Jim Voss) to space station and returned Expedition One crew to Earth.

Mar. 23 **Mir** space station, abandoned in 2000, destroyed during reentry.

Apr. 7 **Mars Odyssey-Delta II** Mars orbiter sent to analyze climate and geologic history and search for water; mission to last more than a Martian year (about 2 Earth years).

2002

Apr. 8 **STS-110** Orbiter *Atlantis* continued assembly of ISS; with two space walks, mission specialist Jerry Ross broke his own U.S. record for most space walks (nine) and total time (58 hrs., 18 mins.).

June 5 **STS-111** Orbiter *Endeavour* delivered Expedition Five crew to ISS and returned Expedition Four crew to Earth; Expedition Four flight engineers Dan Bursch and Carl Walz both broke Shannon Lucid's U.S. space-flight endurance record with 196 days each.

2003

Jan. 16 **STS-107** Orbiter *Columbia,* 113th shuttle flight; at 8:59 a.m. on Feb. 1, radio contact lost during reentry; vehicle broke up at 207,135 feet over East Central Texas, while traveling at 12,500 mph, resulting in loss of vehicle and crew (below, left to right): Dave Brown, Rick Husband, Laurel Clark, Kalpana Chawla, Michael Anderson, Willie McCool, Ilan Ramon.

Columbia crew: NASA

Space Dreams

For millennia, scientists and seers, philosophers and fantasists have speculated about the possibility of escaping Earth's atmosphere and traveling toward the stars. In the 20th century, their imaginings were made real.

The great German astronomer Johannes Kepler, who developed laws of planetary motion still used to chart orbits today, was a dead-serious guy. But upon his death in 1630 he left behind an unpublished manuscript entitled *Somnium,* or *Dream.* The heart of the narrative, which includes a substantial amount of solid science, describes a fantastic voyage to the moon. The most serious guys in the world have always been transported by such flights of fancy.

The ancient Greek story of the inventor Daedalus and his son Icarus is a tale of space travel, climaxing when Icarus flies too close to the sun and solar heat melts the wax holding the lad's feathered wings in place. In the 4th century B.C., the Chinese began making kites and wondering where those kites might take them. In 160 A.D., Lucian of Samosata wrote in *True History,* which may be considered the first work of science fiction, of a ship that sails celestially (although Lucian felt pressed to admit, "I lie . . . I mean to speak not a word of truth"). Firdausi, a Persian poet of the 11th century, spun an epic account of eagles that could transport man into space, and in 1638, Francis Godwin, the Bishop of Hereford, England, employed geese as his avian escorts of choice in *The Man in the Moon: or a Discourse of a Voyage Thither.* The good bishop, no doubt considering his tale less than divinely inspired, was compelled to hide behind a pseudonym, Domingo Gonsales.

Poets and novelists were dreaming things up right along, and so were scientists. The best place to start with serious-minded postulations of the possibilities available in space is the 16th and early 17th centuries, when Leonardo finished his career, Copernicus enjoyed his epiphany, and Kepler and Galileo—among many others—pushed scientific thought forward. Renaissance men all, they were fascinated by art, the human condition in its physical and philosophical aspects, the universe itself.

Culver Pictures

Kepler: Mansell Collection; Galileo: Scala/Art Resource

Among the earliest space casualties was the mythical **Icarus**, whose bravado in approaching the sun proved fatal owing to a technological flaw. Above, Icarus falls from grace in an Albrecht Dürer engraving.

The Italian Leonardo da Vinci was of astonishing acumen as an artist, scientist, inventor and seer. His drawings of the human form seemed divinely inspired, and his depictions of contraptions yet to come were inspiring. Around the year 1500 he made sketches of "ornithopters," flying machines that looked like hybrids of airplanes and helicopters. Many wondered if they might work, and of course one day similar vehicles would.

Between 1508 and 1514, Mikolaj Kopernik, a Polish astronomer known to us as Nicolaus Copernicus, was developing an heretical hypothesis: that the planets orbit the sun—the earth does so once each year—and, furthermore, that the earth turns on an axis once daily. He continued to refine his ideas, and the titanic work that put forth the Copernican Theory in all its revelatory glory was not published until the year of his death, 1543. Kepler, when not writing fictions about trips to the moon, did the much harder work of proving that planetary orbits were elliptical, not round. Solid elements—planets, moons, a star—were being precisely placed within the solar system, and set in proper motion.

As the examples of da Vinci, Copernicus and Kepler show, Galileo Galilei certainly was not the sole engine for moving scientific thought from medieval ideas about natural philosophy to hard, modern theories about the way things work—but he was as important and brilliant as any theoretician, before or since. The former medical student from Pisa became a mathematics professor and a groundbreaking inventor and astronomer. In 1609 he used his homemade telescope to observe the mountains of the moon and, then, details of the Milky Way and deep space never before seen. He openly admitted the inescapable conclusion—Copernicus was right—and for his candor, the Catholic Galileo was, first, admonished by Rome and, later, sentenced to life imprisonment for heresy, since Holy Scripture maintained that God-created Earth was the center of the universe. Galileo would not be pardoned by the Vatican for centuries.

Newton, Descartes and others would add to the growing trove of knowledge about space (Newton's third law of motion, regarding actions having equal and opposite reactions, and his rules regarding gravity, are the underpinning of rocket science). And, in the aftermath of the scientific revolution, those who would posit space travel could lean not simply on pie-in-the-sky notions but also on proven facts, widely held theories and nonfiction science. It is worthwhile, in a confined number of pages, to skip a couple of centuries and numerous books and essays here, while perhaps giving a nod to Joseph Atterlay's 1827 *Voyage to the Moon,* wherein a spaceship travels via antigravity matter called

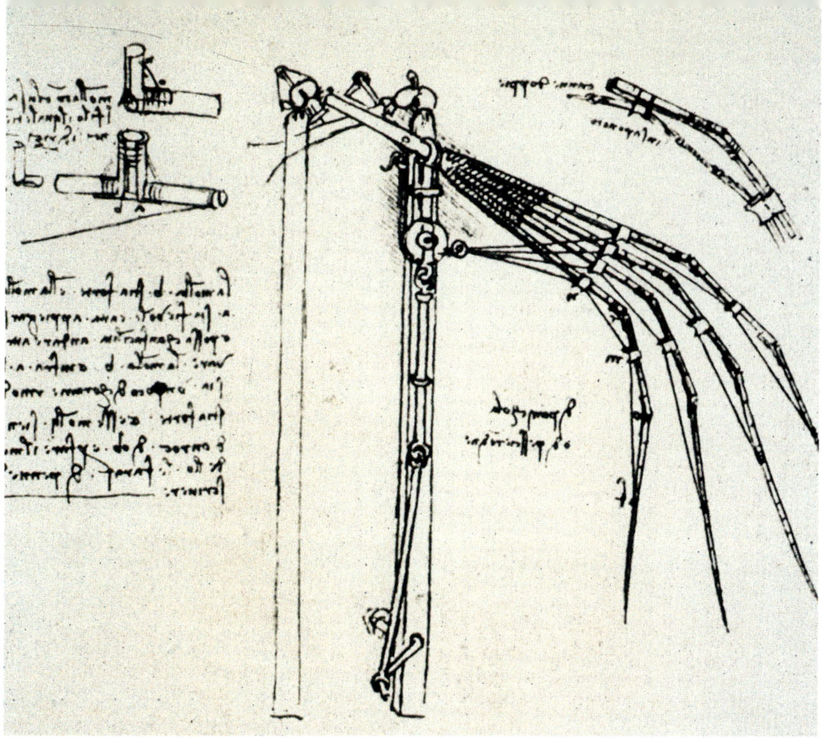

Granger Collection

Leonardo and his designs predicted manned flight, even if they did not go the extra mile and presage space travel. Note the mirror-writing that left-handed Leo employed.

In the 16th century, **Copernicus** changed everything—the way we looked at the heavens, the way we looked at God—when he theorized that the sun, not the earth, was the center of the solar system.

Kepler's work of consequence remains his science, but the tantalizing *Somnium* is as fascinating as anything in his oeuvre. Hidden in it are messages decipherable only by fellow scientists.

When **Galileo** first trained his telescope on the moon in 1609, he opened new highways to the heavens. When he later subscribed to the Copernican Theory, he created his own hell on earth.

Bettmann/Corbis

Lunarium, and to Richard Adams Locke's "moon hoax" of 1835, which, via a series of articles in the New York *Sun,* convinced a large part of the public that lunar creatures had been espied through a supertelescope. Let's skip ahead, and deliver ourselves at Jules Verne.

Among the many fantastic places the French novelist took his vast, rapt readership, the two lastingly associated with him are the bottom of the sea and the realm of the moon. In 1865's *From the Earth to the Moon,* a 900-foot-long cannon propels a manned, 20,000-pound, cone-nosed "moon train" beyond the earth's atmosphere. The adventure continues in 1870's *Around the Moon,* which describes unsuccessful attempts to land on the lunar surface and an eventual splashdown in the Pacific Ocean, where the spacecraft is picked up by the USS *Susquehanna.* Verne understood science and used it in a most convincing manner. It is a measure of his achievement that he was read and respected by working scientists—not to mention countless goggle-eyed young science geeks—throughout the world.

Three modern pioneers who, as adults, freely acknowledged having been influenced by Verne in their youth were the Russian schoolteacher Konstantin Tsiolkovsky, the American physics professor Robert Goddard and the Transylvania-born Hermann Oberth, who would become the driving force of Germany's missile and rocketry program. Each has been dubbed the Father of Space Travel, and each can stake a reasonable claim.

Tsiolkovsky's theoretical work and his science fiction gained little audience in his lifetime but inspired later generations of Russian scientists. His paper, "Exploration of Cosmic Space by Means of Reaction Devices," written in 1898 and published in 1903, outlined the essentials of rocketry, and a quarter century later he was envisioning multistage rockets, artificial satellites and space stations.

Goddard was a hands-on thinker, and in 1926, at precisely the same time that Tsiolkovsky was detailing plans for future orbiting vehicles and stations, Goddard was successfully launching the world's first liquid-propellant rocket. Powered by liquid oxygen and gasoline, the craft zoomed to an altitude of 41 feet above Aunt Effie's cabbage patch in Auburn, Mass., and flew for 2.5 seconds before crashing 184 feet away: a real-world triumph.

Jules Verne belongs in the pantheon not just of sci-fi writers but of space travel pioneers. His ideas had a real impact on real-world rocket men. His leather-padded "moon train" (top) was much cozier than the first capsules. It was roomier, too, fitting three astronauts.

AP

Tsiolkovsky was a poor, deaf teacher all but ignored in his homeland during his lifetime. It was only when other rocketry programs were taking note of his seminal work that the Russians awoke to his good science and remarkable prescience.

Culver Pictures

Robert Goddard, a shy man, was hurt by his public dismissal as "the moon rocket man" after his *Smithsonian* article. He worked all but anonymously through the 1930s and World War II. It is likely that only a lack of government support kept him from equaling or surpassing Germany's V-weapon program. Here, he poses with his liquid-propellant rocket, the world's first, in 1926.

Hank Walker

In 1919, **Hermann Oberth** (center) read a newspaper account of Robert Goddard's work, and thus began his twisty relationship with the United States and its allies. His rocket team in Berlin helped develop the V-2's that rained terror on England in 1944, then his German protégé, **Wernher von Braun** (second from right), led the United States in its race against the Soviet space program. Oberth eventually joined von Braun in the States. This photograph, which includes U.S. Army Gen. H.N. Toftoy (left rear) along with Oberth's fellow expat German rocketeers Ernst Stuhlinger (left) and Robert Lusser (right rear), was taken in 1956.

Oberth was also diligently at work in the 1920s. Born in 1894, he had read *From the Earth to the Moon* a half dozen times in his teens, and had already, in 1917, produced a design for a long-range liquid-fuel rocket; in 1922 he finished a masterly Ph.D. thesis on space travel. But Tsiolkovsky could not get his work published for five years, and Goddard's 1920 *Smithsonian* article "A Method of Reaching Extreme Altitudes" brought the author widespread ridicule. So, too, Oberth's visionary thesis was rejected as too speculative. No matter, Oberth published it in 1923 as a book, *The Rocket into Interplanetary Space.* The volume dealt with orbiting space stations, space suits, space food, space walks and interplanetary travel, and is today considered a classic.

Oberth joined the German Rocket Society in 1927. He set up a team in Berlin that included a young man who had written him a letter upon reading *The Rocket into Interplanetary Space.* Wernher von Braun would later play the lead role in America's nascent space program after World War II, and would, in 1955, bring Oberth to the U.S. for three years' work at the Redstone Arsenal.

All of this speculation and conjecture led to the rockets that fell on London in World War II, and eventually traveled to the moon and beyond.

It is remarkable how narrow the aisle was that separated scientists from fantasists all along the way. In 1928, Hermann Oberth, as important as any figure in putting man in space, was hired as a consultant for the Fritz Lang film, *Woman in the Moon.* Twenty-six years earlier, the pioneering filmmaker George Méliès had already stuck a rocket in the lunar eye in *Voyage to the Moon,* and four decades later, Stanley Kubrick and Arthur C. Clarke would suggest that there might be trouble aboard a probe in 2001. The films of Méliès and Lang were considered entertainments, while critics and audiences speculated that Kubrick's and Clarke's visions might be prophetic. By that time, 1968, man had been in space for several years, and was months away from setting foot on the moon. Thanks to the earlier dreamers, no one was calling them dreamers anymore.

Kobal Collection

In 1902, French filmmaker George Méliès, in a loose, 14-minute adaptation of Verne called **Voyage to the Moon**, had his explorers execute a less-than-soft lunar landing. They were then captured by moon men but escaped.

Fritz Lang, the legendary German director who created the piece *Metropolis,* worked hard to make **Woman in the Moon** as authentic as the knowledge available in 1929 would allow, but got some fundamental things wrong; for instance, a lunar atmosphere. Lang shared the screenwriting with Hermann Oberth, who is credited for "technical material."

2001: Kobal Collection; Moon: Photofest

Writer Arthur C. Clarke first conjured what would become **2001: A Space Odyssey** in his short story "The Sentinel," which he wrote in 1948 and expanded into a novel in the 1960s. The plot concerns a fateful find on the moon, and as *Apollo 11*'s landing module *Eagle* descended toward the lunar surface 15 months after the movie's April 1968 release, *2001* had many wondering, and some worrying.

Donald Uhrbrock, Sovfoto

The Postwar Era

The Race Is On

When the "Reds" launched the world's first spacecraft, Sputnik 1 (above), into what we considered our wild blue yonder, Americans were aghast. If some face-saving was provided by our own Explorer satellite (its rocket fueling at left, as technicians gather blithely nearby), it was nonetheless clear that the gauntlet had been cast.

U.S. Army

Bettmann/Corbis

Von Braun (in cast) and others surrender in 1945. At right, in White Sands, N.M., in '46, von Braun (center) inspects a V-2 engine. When the first V-2 hit London, he told a friend that it had worked perfectly except for landing on the wrong planet.

During World War II, the combatants engaged in a mortal competition to invent ever more deadly means of attack. In Europe, Germany sought to bring the island nation of Britain to her knees by means of an unprecedented horror: destruction that arrived from the heavens with no warning save, perhaps, the eerie buzz of a flying-bomb blitz about to strike home on a random street corner. As the conflict wore on, the battle in the Pacific became one of methodical if costly Allied victories, culminating in the stunning denouement wrought by atomic power.

With the war in its final year, the Nazis, trapped within tightening concentric rings, desperately pursued new methods to strike from afar. The scientists in Germany had for decades shown an acumen for rocketry—but, fortunately for democracy, there was not enough time for them to produce missiles of distant destruction with a potency and accuracy sufficient to seize victory from defeat. Still, by January 1945, they had launched a prototype of the first ICBM, one designed to reach North America by traveling at an altitude of 50 miles and a speed of 2,700 mph. But again, this was only a prototype, and the war's verdict was soon in hand. By the end of the year more than a hundred German rocket scientists and engineers were in White Sands, N.M., having delivered themselves to American troops, if only to avoid being taken by the Soviets.

It would be difficult to overstate the importance of these émigrés to the U.S. space program. American rocketry, under Robert H. Goddard, had been exemplary, as the German star Wernher von Braun was the first to admit: "Until 1936, Goddard was ahead of us all." But post-WWII, the U.S. was thrust into an age of guided missiles, with which it had virtually no hands-on training. Von Braun, who would become chief of the U.S. Army ballistic-weapons program, said of the German rocket men, "If we are good, it's because we've had 15 more years of experience."

The U.S.S.R. also made use of German scientists (some of them kidnapped), and while they played a smaller role, the Soviets feasted on whatever knowledge was available. For them, technology and industry were critical fields of competition in their protracted war with capitalism; furthermore, they were fully aware that American aircraft in Europe could reach their homeland, while the U.S. was yet out of reach. Thus was sounded the communist clarion call for long-range weaponry.

American experiments during and immediately after the war were much more concerned with the effects of sonic speeds on fixed-wing aircraft, piloted by such storied men as Chuck Yeager and Scott Crossfield. Their training was rigorous, their flights perilous in the extreme, but the effort produced a string of enviable successes, and would coalesce with rocketry into a program of destiny.

The Soviets, however, would be a formidable foe in a contest whose brass ring was the universe.

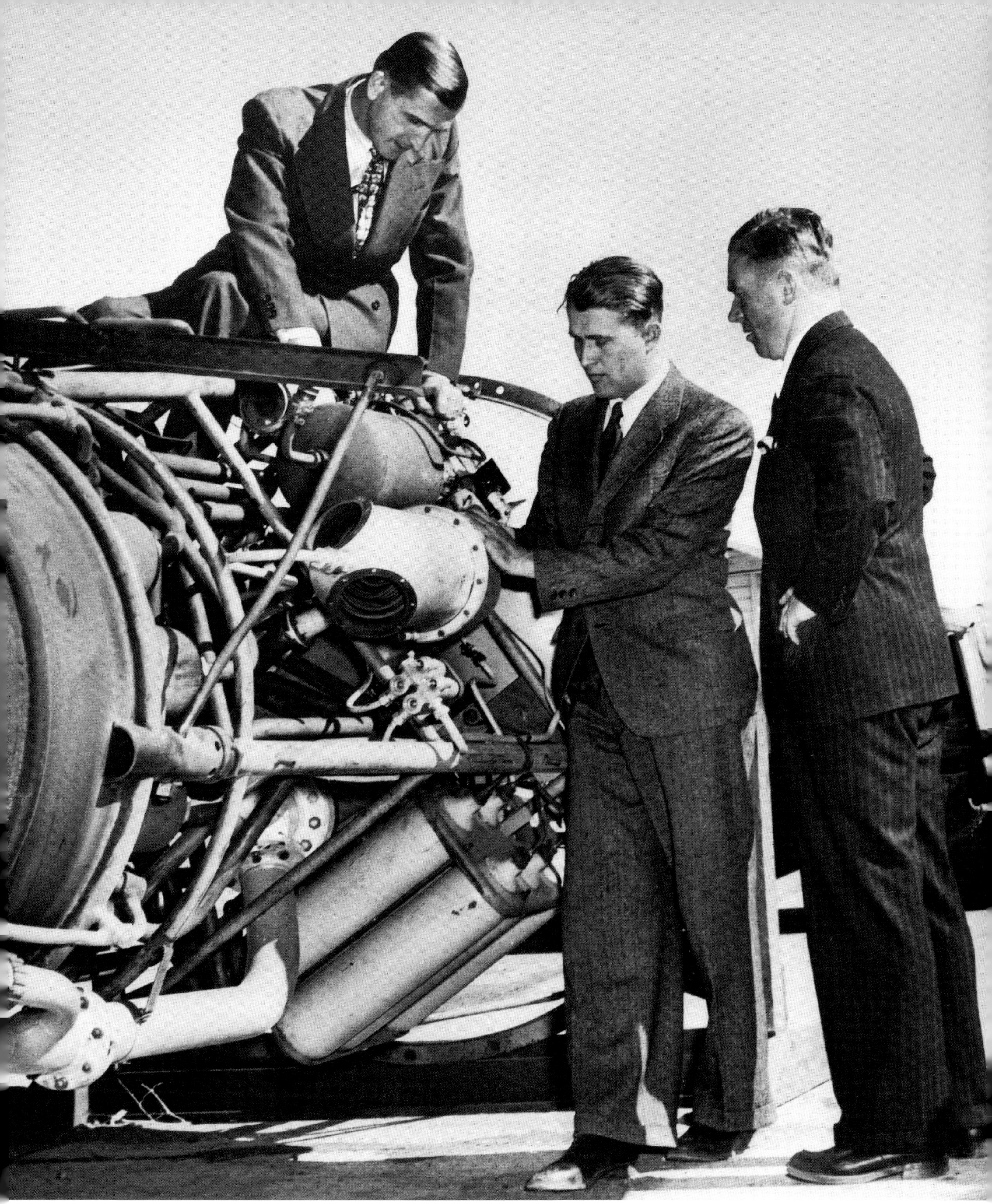

Air Force Test Center, Edmonds AFB

Air Force Test Center, Edmonds AFB (2)

Livin' to Fly, Flyin' to Live

When the hotshot flyboys at Muroc Army Air Base in California's Mojave Desert needed to chill out, they often chose Pancho's. Hostess Pancho Barnes (left) had been a daredevil pilot herself. That's Chuck Yeager leaning over her left shoulder. On October 14, 1947, Yeager and his two broken ribs climbed into a dangerous, rocket-powered beast—an X-1 named after his wife, Glennis—and became the first man to break the sound barrier. At near right, a later version of the X-1. Far right: Yeager's rival Scott Crossfield, the first flier to conquer Mach 2, and the X-15's first pilot, embodies the classic cool and calm of a top test pilot.

Bettmann/Corbis

Allan Grant

Dreams and Nightmares

Their people may not have been living the good life, but the Soviets showed on October 4, 1957, that something was going right behind the Iron Curtain when Sputnik 1 (above, with a streak from its rocket section) became the first artificial satellite. Dithery to catch up, the U.S. often misfired, as with this Vanguard rocket, earthbound moments after liftoff. Its Sputnik-like payload (below) fell lamely into nearby bushes.

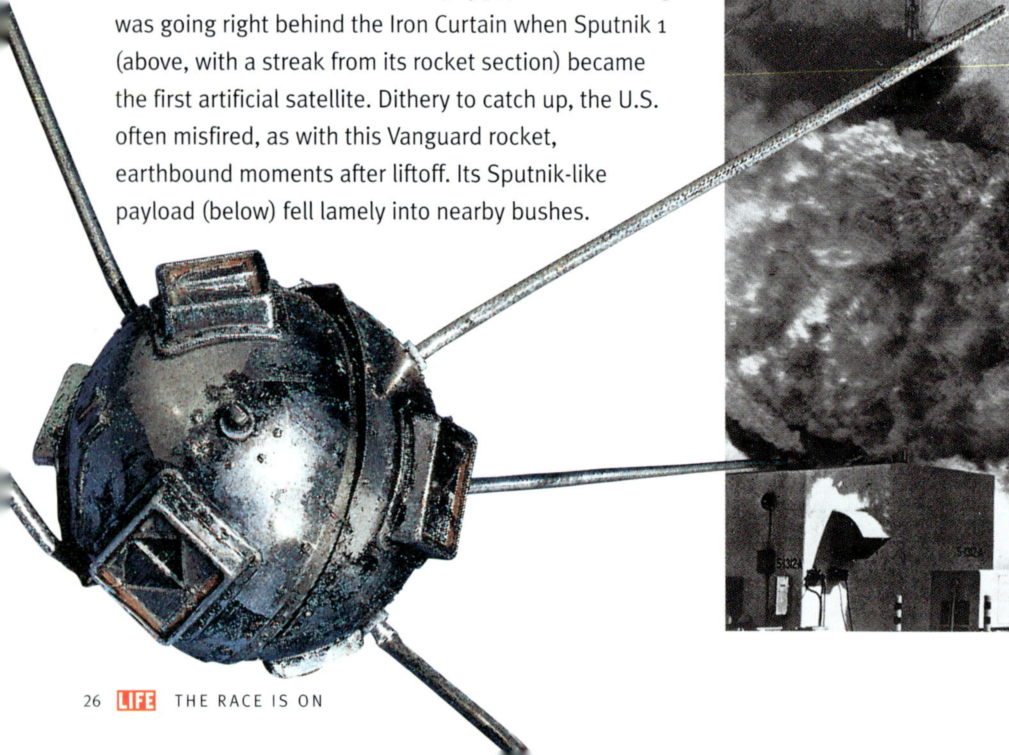

Sputnik: Sovfoto; streak: Robert W. Kelley; Smithsonian (2)

Walter Sanders

U.S. Air Force

The Age of Anxiety
Von Braun, now an American citizen but still a worshiper at the feet of rockets, was distraught that the Soviets had zoomed ahead while the politic U.S. was trying to let the Navy et al. solve a problem that was clearly within his purview. Finally, after the Sputnik shocker, the government allowed von Braun and his Army group to launch the satellite Explorer 1 on January 31, 1958.

Sovfoto

Man's Best Friend

It was bad enough when, just a month after Sputnik 1, the Soviets launched another, much bigger satellite, but then it was revealed that Sputnik 2 had a dog aboard. Now the Soviet plan was clear: They were going to put a man on the moon! Maybe colonize the place! The dog, Laika (Russian for "barker"), became the first of many Earth creatures to "volunteer" for space. Left: The U.S. countered by sending up primates. Right: The Soviet Luna 3 took the first photo of the moon's far side, then LIFE projected it onto a sphere.

Sovfoto

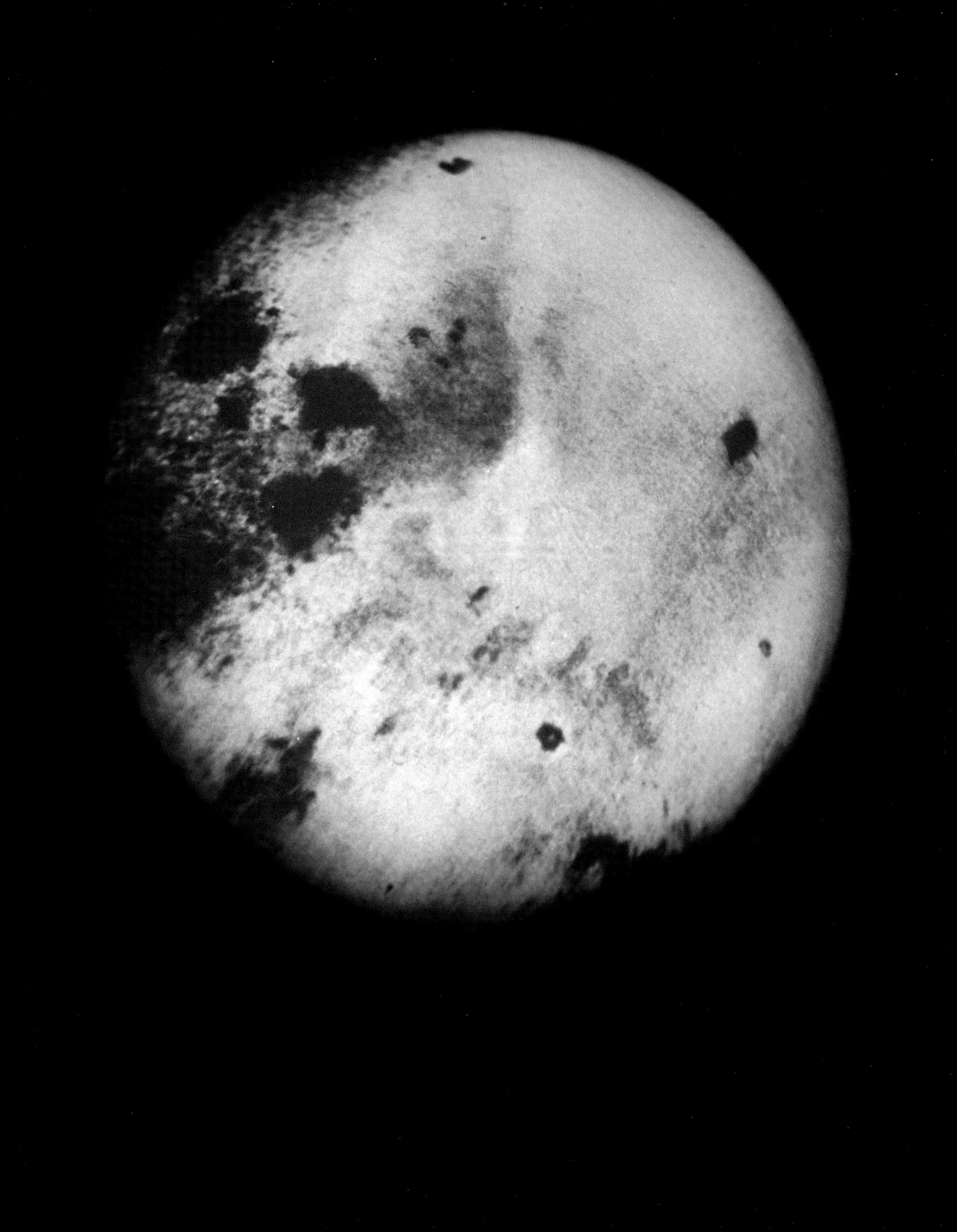

Ralph Morse (2)

Ladies and Gentlemen: the Brave Men (and Women) of Project Mercury

At the end of the 1950s, LIFE registered a great—if controversial—journalistic coup when it agreed to pay the seven Project Mercury astronauts some $70,000 each to showcase their roles. When the first of a series of lengthy articles kicked off (left; the following week's cover showed the wives in the same tableau), the glimpse into the training and homes of America's newest heroes was galvanic, inspiring NASA chief James Webb to thank LIFE for "waking up the country to the space program." Rival publications chafed at the exclusivity of the deal, and NASA itself derided the commercialization of the program—but they had green-lighted it. Finally, even the managing editor of *The Washington Post* conceded that the stories' success relied on hard work more than favored treatment.

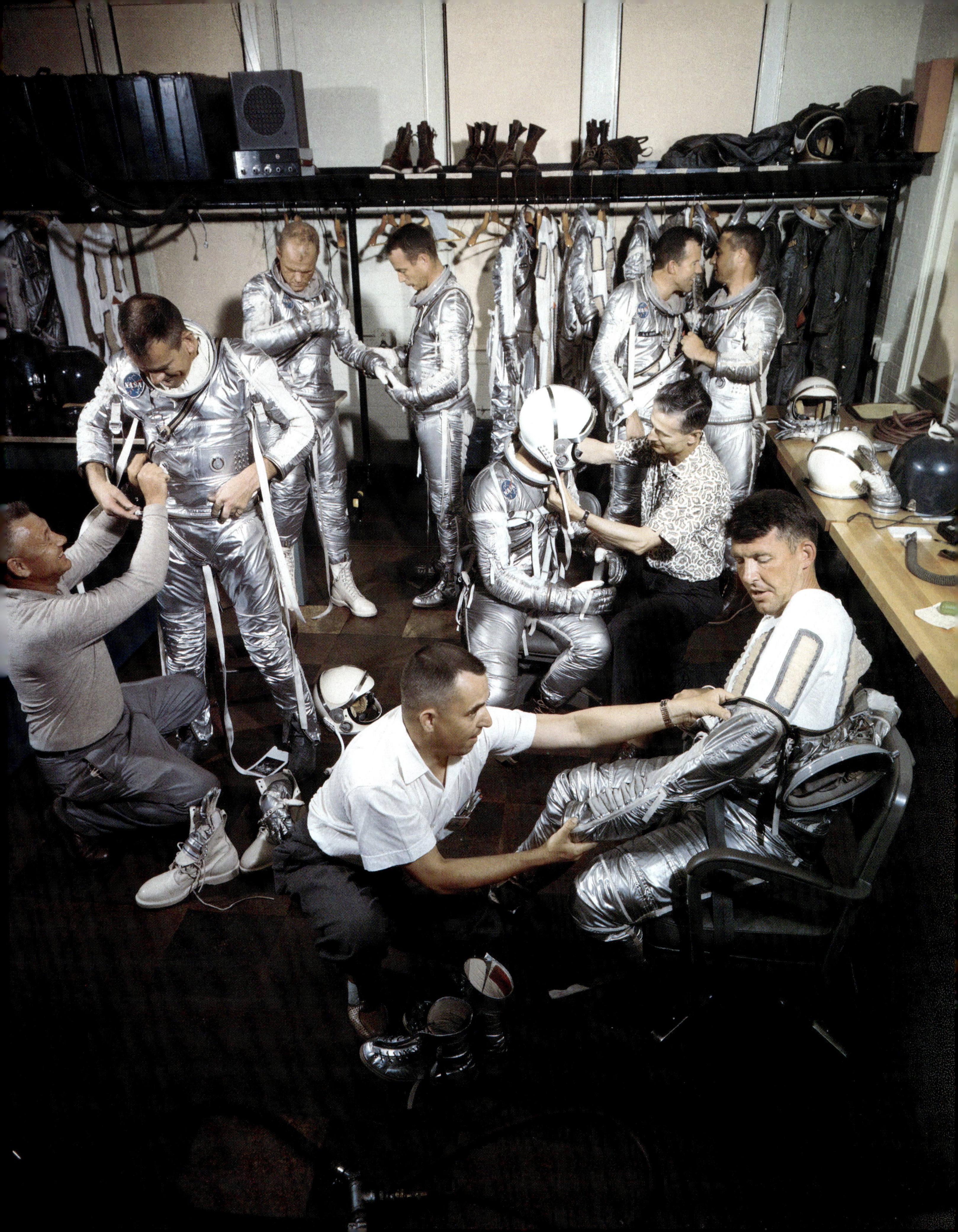

SCHIR.
CARP.
COOP.
GRISS.
SHEP.

Ralph Morse (4)

Only the Strong Survive

Scientists had some idea what it might be like to travel faster than anyone ever had before, to go so high that you were no longer in our planet's atmosphere, to enter a world that had been the domain of fiction. But they also knew there would be surprises aplenty, so they tried to prepare the astronauts for anything. LIFE photographer Ralph Morse spent so much time with the men and their families he almost became a part of the team. Counterclockwise from above: John Glenn prepares for a spin; Wally Schirra at eight g's; Alan Shepard test-drives his personal couch; the team takes a dip.

NASA

Ralph Morse (2)

"Space Voyagers Rarin' to Orbit"

So said the headline in LIFE as the nation got its first look at the seven men who had been picked from a field of 110 candidates. From left, Deke Slayton, Alan Shepard, Wally Schirra, Gus Grissom, John Glenn, Gordon Cooper and Scott Carpenter respond to the query, "Could I ask for a show of hands of how many are confident that they will come back from outer space?" All are in the affirmative, although it may be noted that Schirra shows extra confidence (if also bemusement), while Glenn is basking in bravado. In August of '59 the incessant testing revealed that Slayton had an erratic heart rate. Although he was unable to go up on any of the Mercury missions (above, he watches as an Atlas missile takes off), he was able, with a rigorous effort, to remedy the condition, and in 1975 he took part in the first meeting in space of American astronauts and Soviet cosmonauts.

NASA

“We Should Go to the Moon”

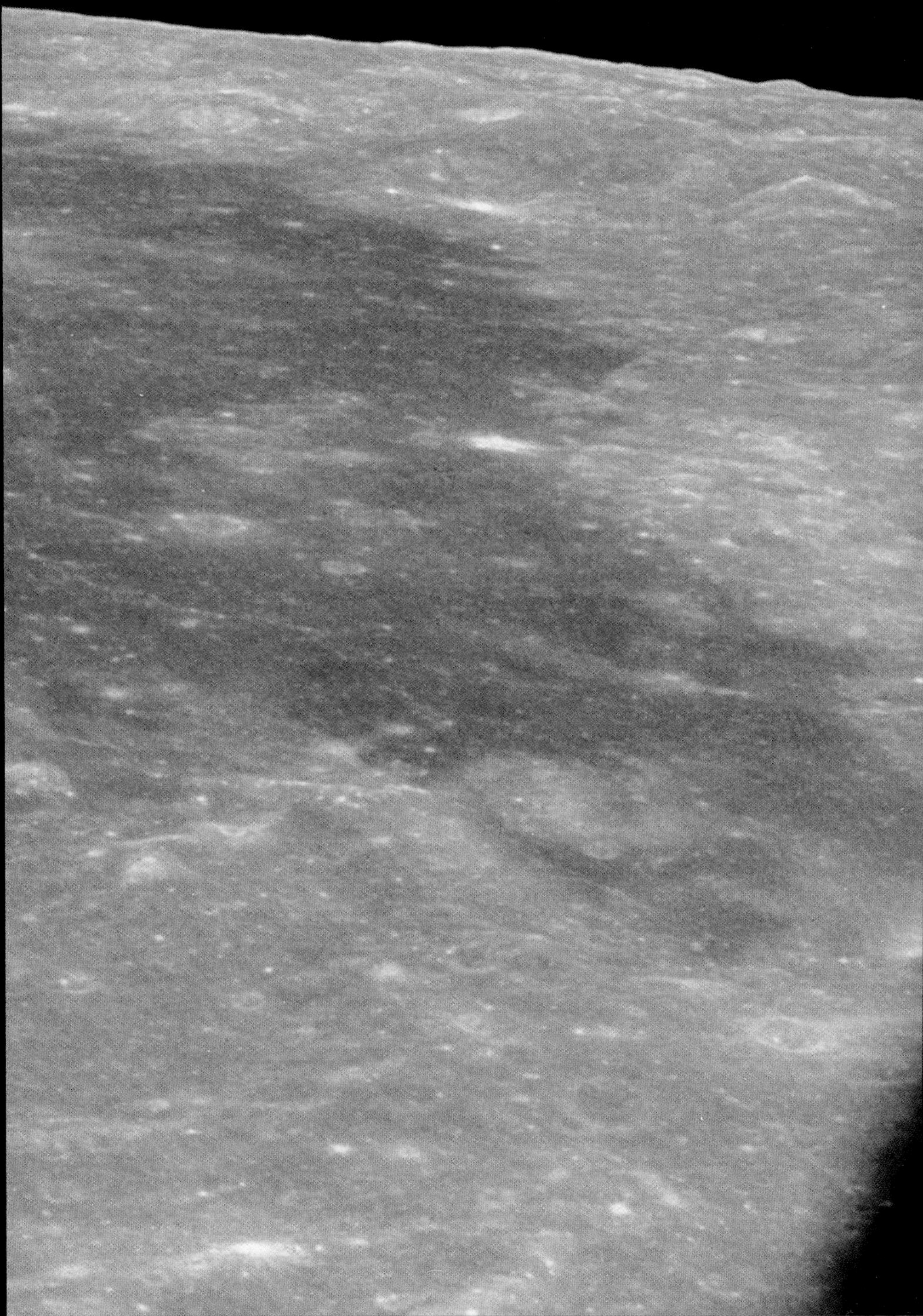

The proposition sounded truly out-there when President Kennedy aired it in 1961, suggesting a deadline of decade’s end. But he saw it as critical to the future of life on earth. The nation rallied behind their leader and cheered as their astronauts set out to catch the Soviets. By the time *Apollo 11* gazed down upon its lunar module in 1969, the Soviet Union seemed clearly visible—in the rearview mirror.

The charismatic JFK had the right stuff for a space race; he even looked like he could have been an astronaut. Opposite: with von Braun. Left: three weeks before the Moon speech, with Alan Shepard, the first American in space. Moments ago, the President had dropped the NASA Distinguished Service Award. Now, pinning it on, he jokes that the medal "has gone from the ground up."

When read today, 42 years after the fact, President John F. Kennedy's Man on the Moon address of May 25, 1961, is a remarkable artifact. The raison d'être for the space race, as we vaguely remember it—that it was about the cold war at least as much as it was about science, adventure, exploration or even conquest—is baldly stated by the President, with no ifs, ands or buts about it. In soliciting billions of dollars for the effort, in "asking the Congress and the country to accept a firm commitment to a new course of action—a course which will last for many years and carry very heavy costs," Kennedy stressed that the outcome of the space race would impact "the minds of men everywhere, who are attempting to make a determination of which road they should take." These roads, said Kennedy, headed in the diametrically opposed directions of "freedom and tyranny." The speech is rife with concerns that the United States is lagging badly, and full of the conviction that, although "no one can predict with certainty what the ultimate meaning will be of mastery of space," this mastery "may hold the key to our future on Earth."

"I believe we should go to the moon," the Chief Executive said. "I believe that this nation should commit itself to achieving the goal, before this decade is out, of landing a man on the moon and returning him safely to the Earth." If the goal itself seemed otherworldly, the rationale behind it was as grandiose as the cosmos. "We go into space because whatever mankind must undertake, free men must fully share." In other words, space simply must not be ceded to the bad guys, and these bad guys are clearly hell-bent on getting there first.

JFK predicted, on that spring day, that the "Soviets with their large rocket engines" would be able to "exploit this lead for some time to come in still more impressive successes." He wasn't wrong. The cosmonauts didn't goof around with up-and-down missions, they went right into orbit. The Soviets sent teams into space while America was still toying around with solo missions, and all of the early breakthroughs—duration records, space walks, first woman in space—belonged to the U.S.S.R.

But something subtle was happening. America's own impressive successes and the nobility and charisma of the strong young men it was boosting aloft caught the fancy of the U.S. public, which grew less concerned with what the other guy was doing. Suddenly—it's hard to say precisely when—the race seemed even. And then, clearly, one program was homing in on its target, a target that had been publicly, brazenly, perhaps incautiously identified on May 25, 1961. "For while we cannot guarantee that we shall one day be first, we can guarantee that any failure to make this effort will be our last," said President Kennedy, pointing to the moon. "We take an additional risk by making it in full view of the world, but as shown by the feat of astronaut Shepard, this very risk enhances our stature when we are successful."

Bob Gomel

Ralph Morse (3)

Flies on the Wall, or in the Ointment?
With LIFE photographer Morse and his colleagues buzzing about, the Mercury 7 were rarely out of focus. Some were more adept than others at striking the nonchalant pose. Scott and Rene Carpenter, at home in Virginia, were fine; Glenn, riveted by a rocket test along with Grissom, Shepard and family members, was terrific; Shepard, after a jog, wasn't so hot. And although happy to cash LIFE's checks, the guys weren't keen on full disclosure, as Tom Wolfe reported in *The Right Stuff:* "Oh, the LIFE writers sometimes had their own notions of what was candid and colorful and 'good copy.' They liked to get on such subjects as the rivalries between the boys and such 'colorful' matters as Drinking & Driving and the unspoken intrafraternal business of fear and courage . . . Well, the hell with that! It was not so much that the men wanted to come out sounding like the Hardy Boys in Outer Space—it was just that you'd have to be an idiot to let your personal story actually get personal."

TASS/Sovfoto

The First Man in Space

On April 12, 1961, the Soviet Union stunned the world when cosmonaut Yuri Gagarin, riding the spacecraft *Vostok 1* (left), returned safely after an orbit of the earth. Later in April, Gagarin walks the red carpet at Moscow Airport, then thousands cheer him in Red Square as Yurimania sweeps the steppes. Paranoia, gloom and doom settled over the American landscape. There were, it should be noted, small glimmers of détente during the tense space race. Gagarin visited America even as the cold war raged, and in one sweet gesture, President Kennedy was given a puppy from a Soviet space dog's litter. Aides worried that microphones had been hidden inside the pup, but JFK would not allow wee Pushinka to be put down. The little cutie eventually had offspring of her own, sired by the White House pooch, Charlie. JFK dubbed them pupniks.

James Whitmore (2)

Слава советской Родине!

Dean Conger

Ralph Morse (2)

Alan Shepard's 15 Minutes

The Navy commander from New Hampshire was chosen as one of three finalists—with Glenn and Grissom—for the inaugural flight, before being anointed the Man. If he looks like he is sprinting to the Redstone booster rocket in the predawn hours of May 5, 1961, then looks hardly deceive: Ever since Gagarin's triumph, the

operative phrase within NASA's Space Task Group had been, "Let's go!" John Glenn, having checked out the Mercury capsule that sits atop the Redstone, gives Shepard an A-O.K. Shortly thereafter, some 45 million viewers were glued to their television sets for their first experience of what would become an American ritual in the '60s. At 9:34 a.m., there was liftoff, and 2.3 seconds later came Shepard's voice: "Ahh, roger . . . Yes, sir, reading you loud and clear. This is *Freedom 7*. The fuel is go; 1.2 g; cabin at 14 psi; oxygen is go . . . *Freedom 7* is still go." Shepard did not orbit. In fact, his flight, which lasted only 15 minutes and 22 seconds, splashed down only 302 miles from Cape Canaveral. But it had put America in the race, and Alan Shepard in the history books.

NASA (3)

U.S. Navy/NASA

Dean Conger/NASA

"It Was a Good Flight and a Great Float"

That was the headline over an exclusive story in the July 28, 1961, LIFE, a first-person account "by Virgil Grissom." About half of the statement was true: Grissom's 15-minute, 37-second flight from Florida to the Atlantic had been a near replica of Shepard's: smooth liftoff, five minutes of weightlessness, a near-bull's-eye splashdown. But then the sailing became a little choppy. Grissom said that the capsule's hatch blew open by itself before the recovery helicopter latched on. Others, including some NASA engineers at the time, wondered if the pilot had panicked and triggered the hatch too early. Whatever: The capsule filled with water and sank, Grissom was plucked from the ocean by a second helicopter, and an heroic face was painted over the whole ordeal—with Grissom, NASA, President Kennedy and, yes, LIFE in willing complicity.

The Right Man for the Job

There is no question that John Glenn, the Marine ace who had distinguished himself in WWII and Korea, desperately wanted to be the first American in space. As it happened, he was the third but proved to be the perfect pilot when it came time for an American to orbit the earth. After Grissom's mission, NASA turned its efforts to extended flight, sending an unmanned Mercury-Atlas combo into orbit, then one captained by the chimp Enos. On February 20, 1962, Glenn clambered into his *Friendship 7* capsule and was launched into orbital flight. As with NASA's previous manned mission, there were snafus, though in this case the astronaut saved the day. After Glenn's autopilot failed, he steered the capsule himself during the final two of three orbits. When a loose heat shield was detected, Mission Control ordered changes in the reentry procedure—which were made, and Glenn stoically rode out the storm.

Lynn Pelham

Michael Rougier

Ralph Morse (2)

Hail the Conquering Hero!

As Annie Glenn and her children get the news that Dad is safe and sound, their relief is palpable; later, with Vice President Lyndon Johnson as chaperon, John gives Annie a kiss. While merely a fifth the duration of the most recent Soviet mission, Glenn's four hours and 54 minutes circling the planet sent the country into a pyrotechnic display of patriotic pride. NASA was flooded with requests for its golden boy, and resorted to sending surrogate astronauts since Glenn couldn't be—quite—everywhere at once. The Real Deal did appear before a joint session of Congress (where he would wind up spending time from 1974 to '99 as a Democratic senator from Ohio) and in several ticker-tape parades, including the biggie in New York City (right). Three more Mercury flights in 1962 and '63, piloted by Carpenter, Schirra and Cooper—who orbited 22 times in 34 hours—closed out the program in fine style.

Sovfoto (2)

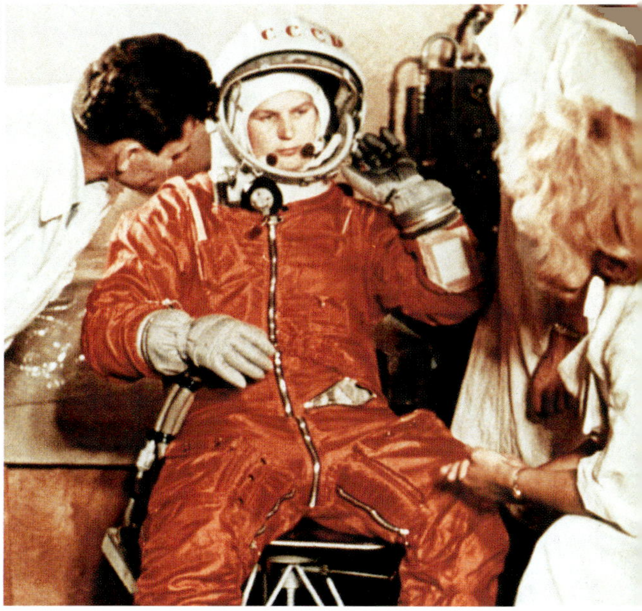

Meanwhile, in Moscow . . .
Although the Yanks were appearing to hit their stride, the Soviets continued to set the pace. On June 16, 1963, a month after Gordon Cooper's long ride, the U.S.S.R. sends Valentina Tereshkova (above) aloft for nearly three days, making her the first woman in space. Her *Vostok 6,* part of a dual mission, passed within three miles of the *Vostok 5* capsule. In March 1965, Alexei Leonov spends 10 minutes outside *Voshkod 2:* the first space walk.

NASA (3)

"I Can Sit Out Here and See the Whole California Coast"

The purpose of the early space walks was to test techniques that future astronauts would use in docking or while making repairs, but to judge by the experience of America's first walker, Ed White, the strolls were a hoot. During his 17,500-mph perambulation on June 3, 1965, White—120 miles above the Pacific—cavorted and laughed. When his partner in *Gemini 4,* Jim McDivitt, ordered him back in after 15 minutes, White said, "I'm not coming . . . This is fun." Finally giving in, White pronounced it "the saddest moment of my life." As McDivitt shot photos from the cockpit, two rapid-sequence cameras on board captured the images below.

NASA (4)

Rendezvous and Docking

In scenes that must have informed the vision of director Stanley Kubrick, the several spacecraft of the Gemini program moved in and out, up and over, while occasionally discharging astronauts into space, as they conducted orbital maneuvers in 1965 and '66. Above: *Gemini 6* and *7* (the approaching ship), both of which are manned, cozy up in December 1965 for the first rendezvous in space; Frank Borman and Jim Lovell will spend two weeks aboard *Gemini 7* and complete a record 206 orbits. Left: In July 1966, *Gemini 10*'s John Young and Mike Collins peer down at an unmanned Agena rocket, with which they will shortly dock. The Gemini exercises represented crucial steps toward a lunar landing and, later, a space station.

No Sweat/Considerable Sweat

A Navy frogman jumps from a helicopter to secure a flotation collar around the *Gemini 5* capsule, which has splashed down after carrying Gordon Cooper and Pete Conrad on the first extended manned flight: seven days, 22 hours. (The sea is green with target dye released from the spacecraft.) Neil Armstrong and Dave Scott appear serene in their *Gemini 8* capsule on March 16, 1966, but they have been through a harrowing ordeal. After they had docked with an unmanned Agena rocket, the combined unit went into an unexpected spin. Deft maneuvering by the crew freed *Gemini,* but 70 percent of its fuel had been spent. The mission was curtailed, and Armstrong brought the ship home.

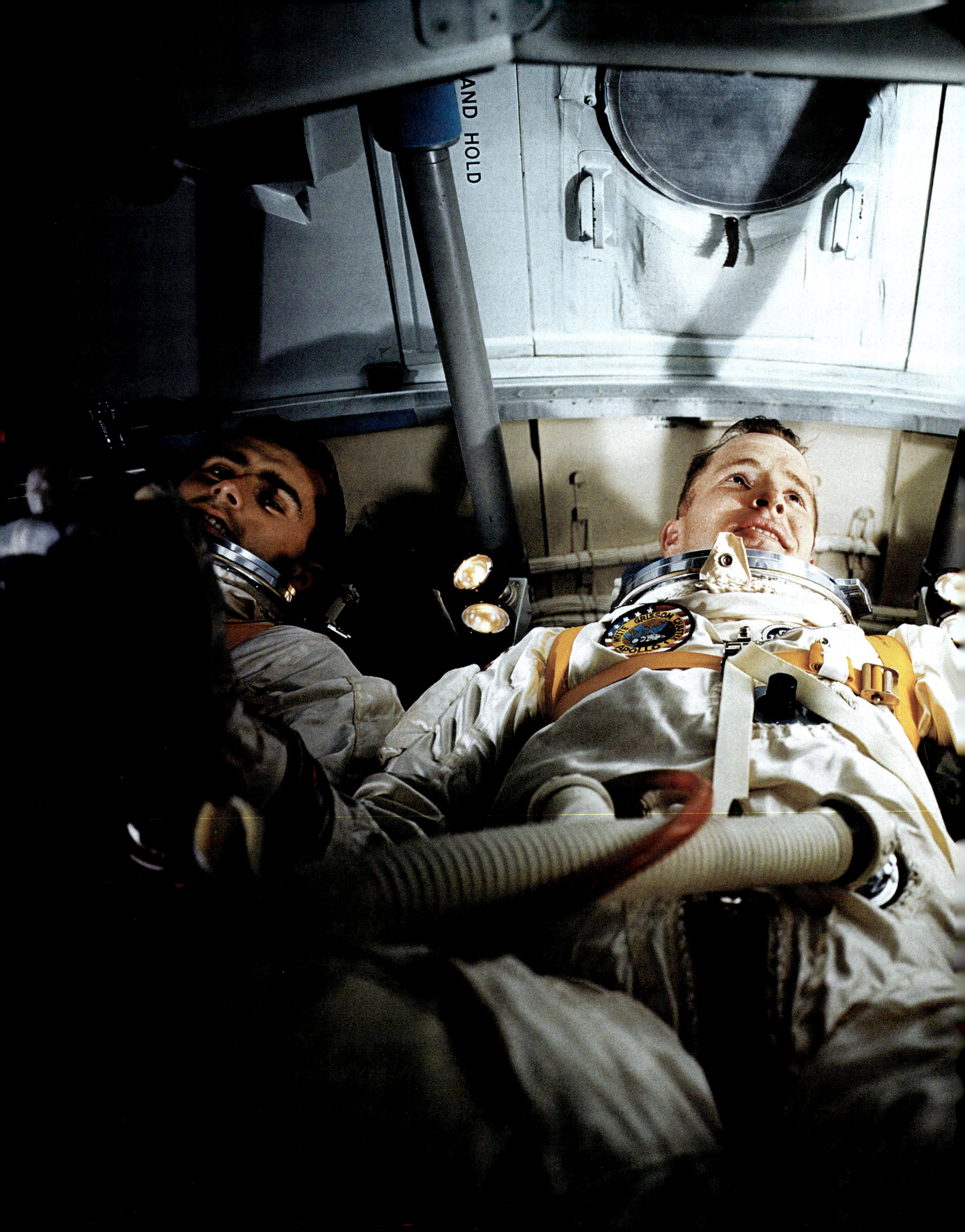
AND HOLD
WHITE GRISSOM CHAFFEE
APOLLO

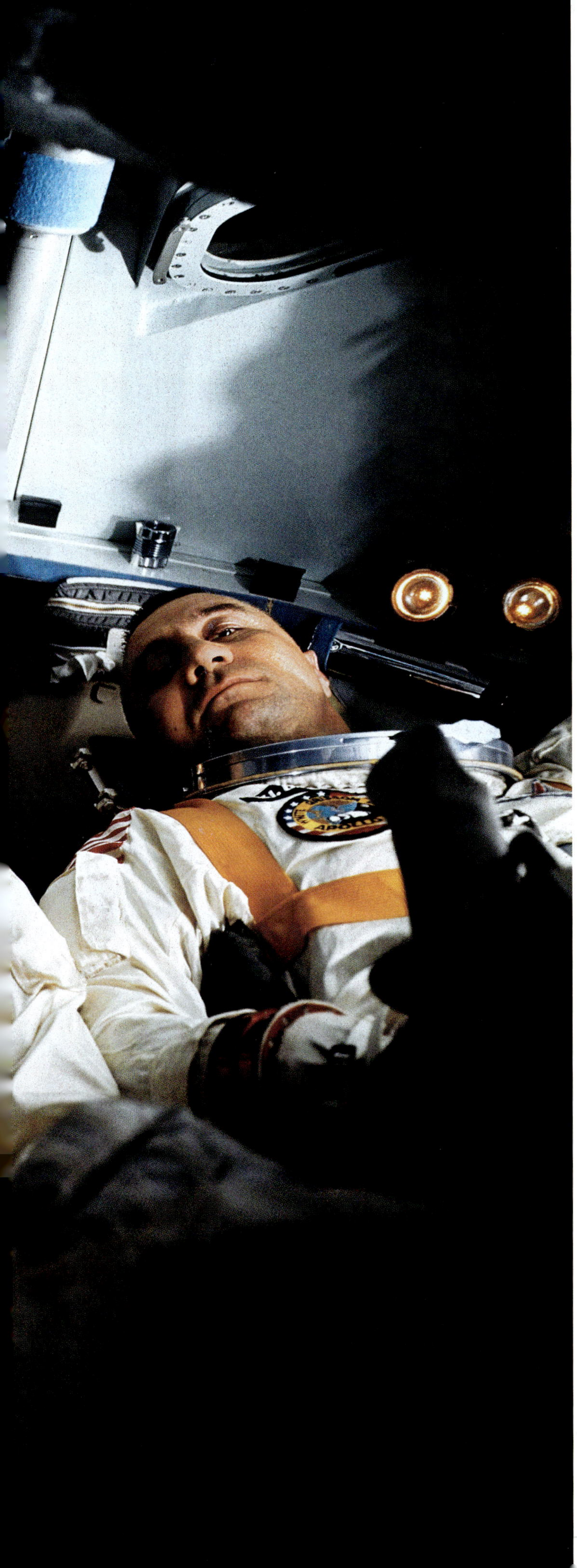

Ralph Morse

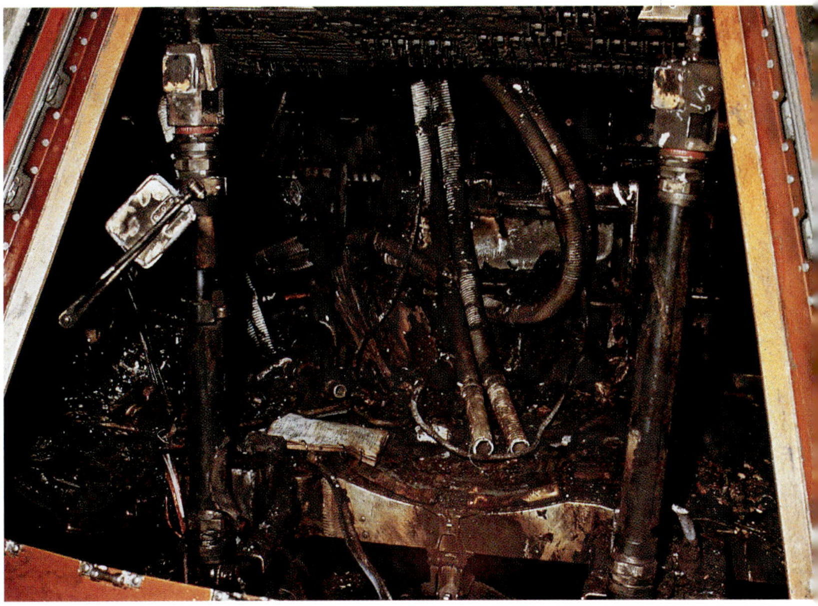

NASA

First Tastes of Tragedy

In a speech to the U.N. in 1963, President Kennedy called space travel "the most hazardous and dangerous and greatest adventure on which man has ever embarked." Several years without a human fatality lulled us into a sense of security, but then, on January 27, 1967, a short circuit during a mock launch sequence aboard *Apollo 1* set off a flash fire that killed (from left) Ed White, Roger Chaffee and Gus Grissom. Below: Fellow Mercury veterans (from left) Carpenter, Glenn and Cooper attend Grissom's casket as taps drifts over Arlington cemetery. Only three months later, Vladimir Komarov died when his capsule's parachute lines tangled in reentry. He, Gagarin and other cosmonauts had protested that the new Soyuz spacecraft wasn't yet flightworthy, but Premier Leonid Brezhnev ordered the launch. Komarov was still cursing as he crashed. A rumor maintains that when Gagarin next saw Brezhnev, he threw a drink in his face.

Paul Schutzer

NASA (2)

Vernon Merritt

Back on Track, with the Moon in Sight

NASA's programs—Mercury, Gemini, Apollo—were in development simultaneously; work on an Apollo craft had begun in 1961, and flight testing in '64. The launchpad disaster that claimed three lives stalled operations for 20 months, but in October 1968, *Apollo 7*, the first manned Apollo mission, was sent forth. Above: *7*'s Saturn I-B rocket seems to blast off right next to the "barn" (then the world's second-largest building), where huge new Saturn V's are being assembled. The photo, taken from 35,000 feet, is an optical illusion: The building is five miles from the launch site. Astronaut Donn Eisele's wife, Harriet, frets as the rocket lifts off, but all goes well during the 11-day mission. Opposite: Dave Scott gazes from the hatch of the *Apollo 9* command module toward the yet-untested lunar module. Later, colleagues Jim McDivitt and Rusty Schweickart will fly the little bug, code-named Spider, as far as 100 miles from the mother ship.

CAUTION
CAUTION

Ralph Morse (3)

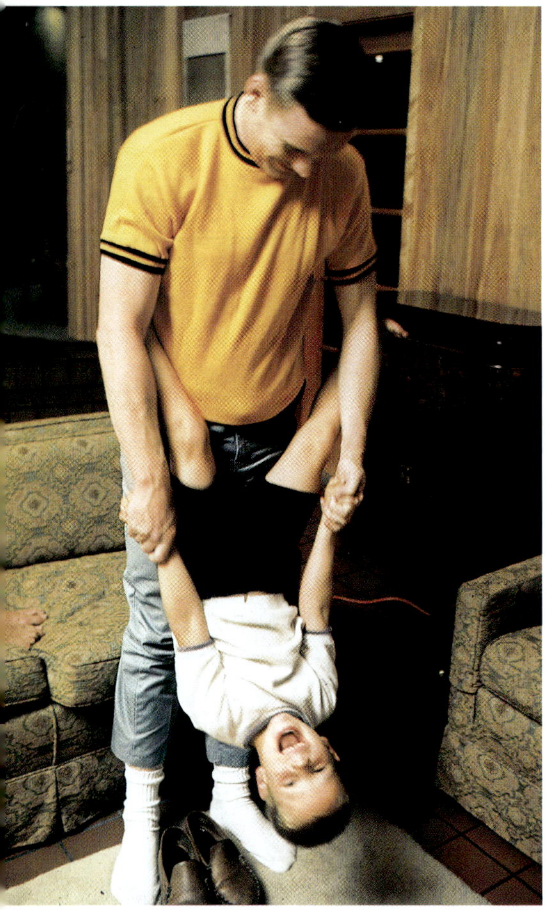

Lee Balterman

The Chosen Ones

In January 1969, Joan Aldrin started to sense that something was up when her husband, Buzz, began acting withdrawn. When she learned that he had been selected for the *Apollo 11* moon-landing crew, she didn't know whether to cheer or cry. "I want him to do what he wants," she told LIFE, "but I don't want him to." So it was for all the family members of these three space veterans, whose new assignment was to go to the moon and return to Earth, with two of them setting foot on the lunar surface. In the weeks leading to liftoff, there was nothing but quality time in and around the modest Houston homes where Neil Armstrong, Mike Collins and Aldrin lived. Counterclockwise from above: Clan Collins goes riding; Armstrong dangles son Mark; Aldrin hangs with son Mike; Aldrin simulates a flight path for partners Collins and Armstrong.

Ralph Morse (4)

The Stuff of Dreams

It seemed suitable that, long before the liftoff of *Apollo 11,* Cape Canaveral had been renamed Cape Kennedy, in honor of the late President who had energized the space program and set the moon as its lofty goal. (The Cape reverted to its Spanish name, which means "place of reeds," in 1973.) Before dawn illuminated July 16, 1969, half a million spectators and more than 3,000 journalists filled the nooks and crannies around the space center to watch the historic event. Armstrong led his team from the gantry to the massive Saturn V rocket that would boost his command capsule, *Columbia,* into space. At 9:32 a.m., the ground shook when 7.5 million pounds of thrust powered the 3,050-ton *Apollo*-Saturn stack off Pad 39A. (LIFE's Ralph Morse operated a rapid-fire camera from a remote location to record the sequence above.) In short order, the spacecraft broke free of Earth's gravity, then spent three days plying space on its way to lunar orbit. There, its fragile module bearing two astronauts would be released, and would aim for yet another aptly named site: the Sea of Tranquility.

Ralph Crane

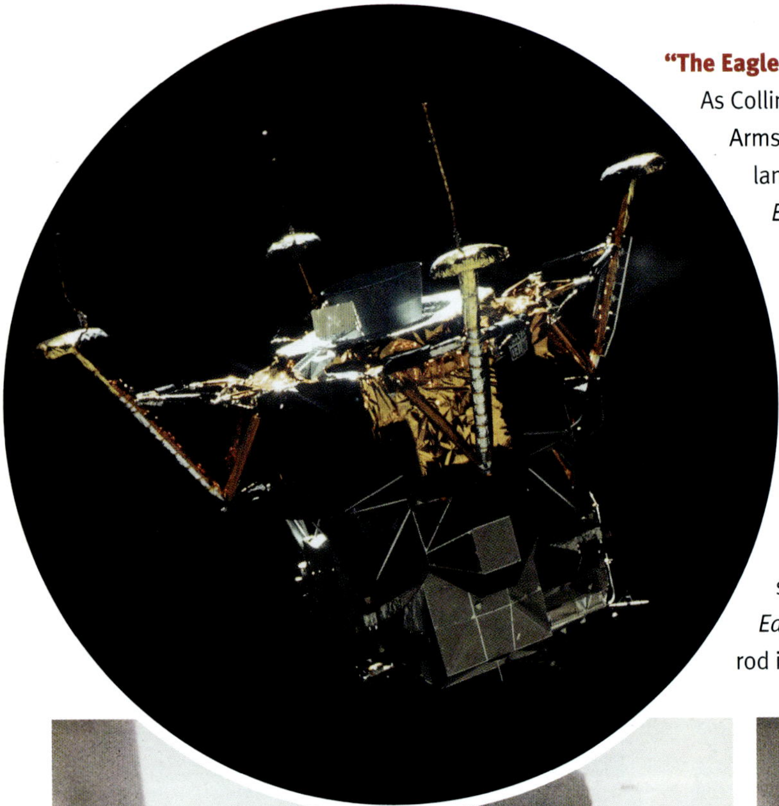

"The Eagle Has Landed"

As Collins orbits overhead in the *Apollo* command module, *Columbia,* Armstrong and Aldrin execute a "powered let-down" in the lunar lander *Eagle* (left). The descent was unnervingly suspenseful: *Eagle*'s computer kept ringing false alarms, Armstrong had to take manual control, the quarter-million-mile radio transmission kept crackling, and the craft alit with less than a minute's worth of fuel remaining. At last, *Eagle* reported, "Houston, Tranquility Base here. The *Eagle* has landed." To which Houston replied, "Roger, Tranquility, we copy you on the ground. You got a bunch of guys about to turn blue. We're breathing again. Thanks a lot." Below: Armstrong departs the module first, famously calling his progress, "One small step for man, one giant leap for mankind." (He later clarifies this as "One small step for *a* man.") Opposite: Aldrin's gold-plated visor mirrors *Eagle* and Armstrong, who is taking the picture; the foil-wrapped rod in front of Aldrin is an *Eagle* landing probe, bent at touchdown.

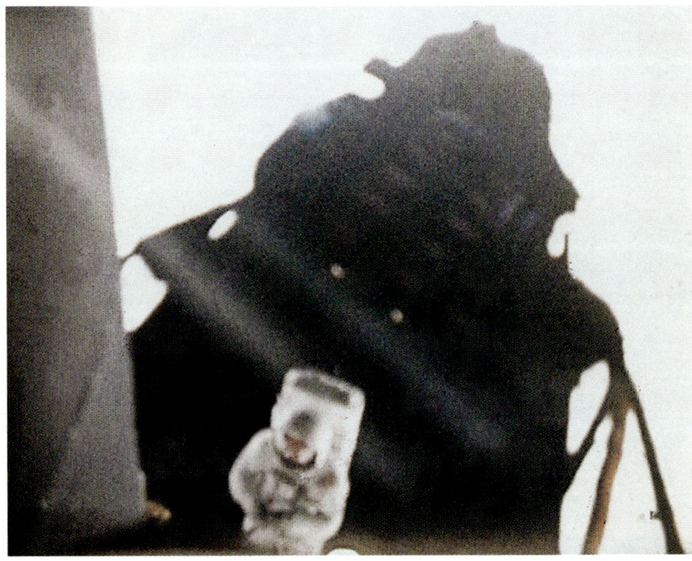

NASA (6)

NASA (2)

A Mission of Exploration and Science, but Not of Conquest

On the evening of July 20, 1969—the way these fellows would have reckoned it back home—Armstrong and Aldrin spent more than two hours walking on the moon. They laid out scientific experiments, setting up a laser reflector, and also a seismometer to measure any moonquakes. At right, Aldrin unfurls the "solar wind sheet," designed to collect tiny particles blowing from the distant sun. The men gathered lunar samples for later analysis. They also turned their attention to matters cosmological, in a nonscientific sense of the word. They planted an American flag, stiffened with wire so that it would "wave" in the lunar atmosphere (above, seen from the lunar module, just before liftoff). They were, emphatically, not claiming the land for the United States, and they also left behind a plaque with the inscription, WE COME IN PEACE FOR ALL MANKIND.

NASA

Bob Peterson

Vernon Merritt

Jubilation

Falling within JFK's lunar-landing deadline by nearly half a year, the achievements of *Apollo 11* brought back all the thrill and wonder of the earliest Mercury missions—and then some. About 600 million people in 47 nations watched the first step, and the reaction, everywhere but Red Square, was of earthly joy. For the families, splashdown brought relief in equal measure with pride. Clockwise from right: Jan Armstrong (center, with Jim Lovell) beams; Joan Aldrin applauds as she watches Buzz struggle aboard the raft; Pat Collins (in red) and friends toast the astronauts with bubbly; President Richard M. Nixon greets the men in quarantine aboard the USS *Hornet*.

John Olson

NASA

The 1970s

New Vistas and High Drama

If sending a man to the moon was straight out of Sci-Fi 101, then launching a space station was just as hoary, though the subject of a more advanced course. The notion, however, of partying in space with an "enemy" seemed entirely original. Here, the team of *Skylab 1* approaches the lab and sees firsthand the damage. Those broken cables on top are all that remain of a missing wing.

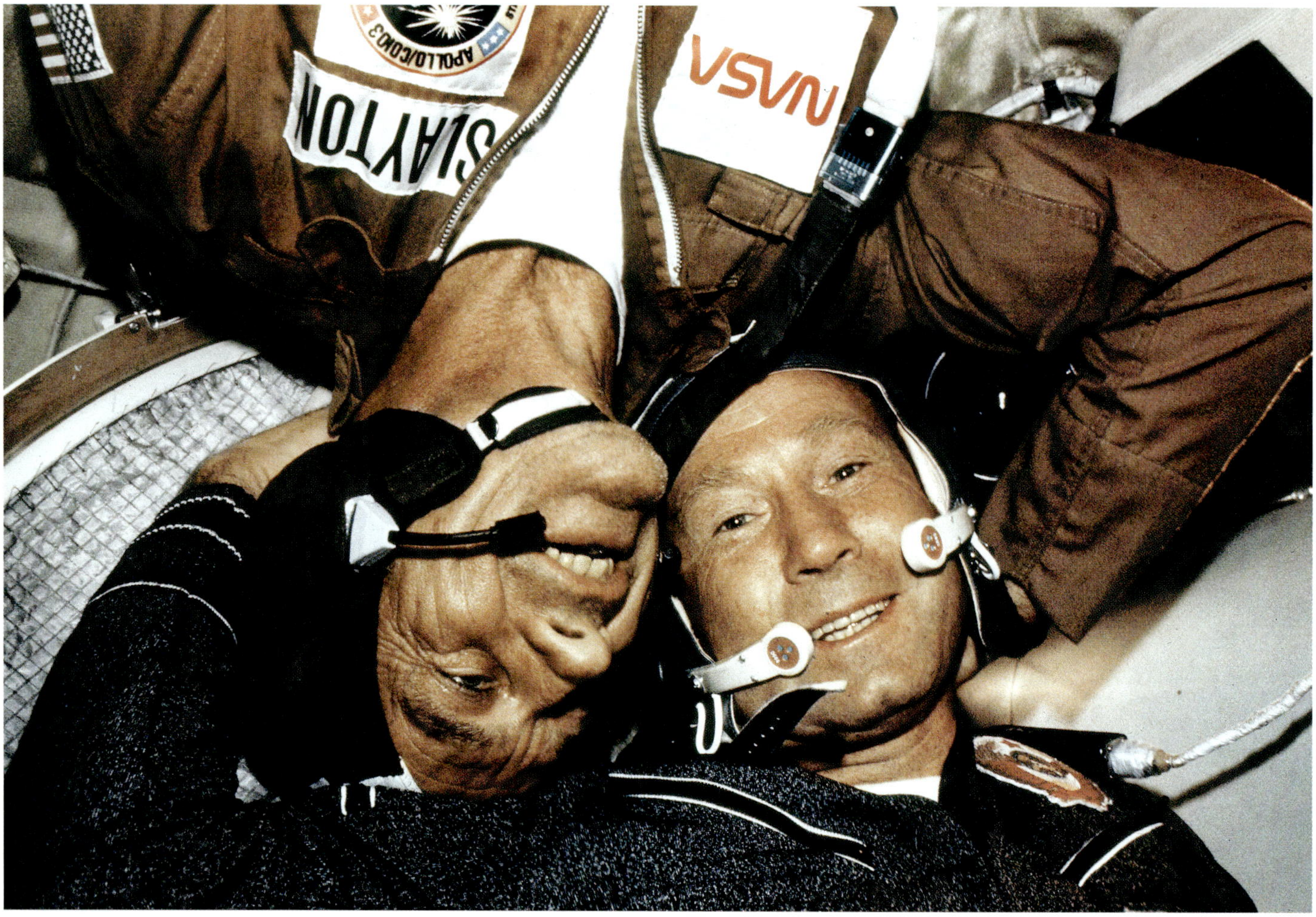

NASA (2)

After centuries of conjecture and dream, mankind had taken a giant leap, then another, and was now eager and able in this new decade to proceed with vigor in the attempt to tame outer space—and, perhaps, in the offing, outdo the other fellow.

If life in space for American astronauts was rocketing along gloriously, the scenario at home was less sanguine. The United States was in terrible turmoil as the decade opened, a society cleft by social and intellectual differences. The bloody specter of the war in Southeast Asia loomed larger than ever, the economy was headed for a serious recession marked by a wage and price freeze, and the numbers on crime were more than disquieting. As he had been for Neil Armstrong's hop onto the moon, Richard Nixon was President. In an interview with *Time*'s Hugh Sidey, he later said, "Just think how miserable it would have been had we not had the space success when we were in the midst of Vietnam, then Watergate and all that."

Indeed, "all that" was a heavy burden, a test, even, of the nation's very stability. But although the astronauts were the subject of a singular celebrity, the stars of stunning ticker-tape parades, few of us realized at the time how important they were to maintaining the social fabric by means of scientific warp and social weft. Even to those citizens who might have considered the money allotted to the space program a misplacement of funds, the epic scale of the events and heroic stage on which these astronauts performed were undeniable. These were mortals who walked among us, and on the moon. Who knows where they might tread next? What dramas might ensue?

On a less grand level there was one certainty—the cold war, which trudged ever forward, and political relations between the two superpowers suggested little reason for optimism. In the realm of science, however, there was proof that the two space rivals could work together. "Good show!" exclaimed the skipper of the *Soyuz,* after the spacecraft "shook hands" with *Apollo.* It was but a small step for these old foes, but surely this distant amity was a brick in the foundation of future cooperation, if not peace, on Earth.

Exploration in the 1970s embraced both the sublime and the frightening. Astronaut Deke Slayton and cosmonaut Alexei Leonov frolic as the U.S. and the U.S.S.R. hook up for the first time in the thrilling frontier of outer space. Opposite: The troubled service module for *Apollo 13,* separated from the command module, drifts in the cold, dark austerity of space.

NASA

Bill Eppridge

Everything to Fear

Science doesn't abide superstition, however tempting. Still, there are those who might have predicted trouble for this 13th Apollo, which was slated to take off in the 13th minute after the 13th hour. Then, two days in, on the 13th day of April 1970, crisis. The world was alerted to the plight of Jim Lovell, Fred Haise and Jack Swigert when Swigert said, "Houston, we've had a problem here." An explosion had ruptured an oxygen tank, leading to the loss of air, water, power and propulsion on the service module. The crew shifted to the lunar module, where the heat was shut off to conserve juice (left, Lovell in the chill). Above, Marilyn Lovell slumps over a squawk box in her home, with her husband so many miles away.

NASA (3)

A World Awaits Their Fate

Across the globe, people were glued to the drama of the three men who might not return. After four long days, and expert course corrections by the gutty astronauts, the command module (top left) approaches splashdown in the South Pacific, watched by perhaps the largest audience in TV history to that time. Above, Mission Control relaxes after touchdown. In the foreground, 36-year-old flight director Gene Kranz fires up a cigar as Lovell talks on the screen at rear. Left: Haise, Swigert and Lovell are welcomed aboard the USS *Iwo Jima* by recovery task force commander Rear Adm. Donald C. Davis. The efforts of the three were lauded everywhere, though, as Jack's father, Dr. J. Leonard Swigert, said, "It was a wonderful beginning and a beautiful landing. But I wouldn't give you two hoots for the interim."

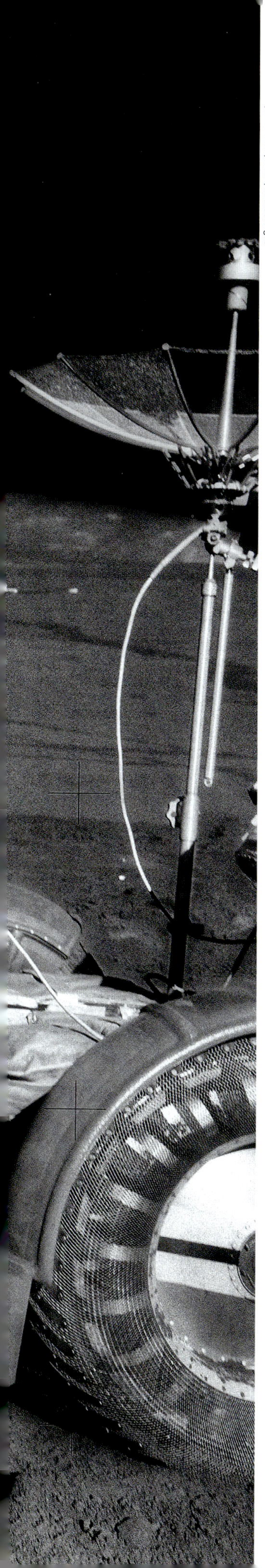

James Irwin/NASA/Michael Light

NASA (2)

Back on the Horse, er, Buggy

Following the *Apollo 13* scare, the next, and final, four missions of the program went off pretty much as planned and without any attendant frightfest. Above, with *14*, Alan Shepard, the first American in space, again took center stage, shielding his eyes from the sun in this photo taken from the lunar module. At right, the three-quarter-inch-deep track of the rickshaw-like transport that the 47-year-old Shepard and his colleague Edgar Mitchell used for ferrying geological tools and collecting rock samples. Left: *Apollo 15* Commander Dave Scott test-drives the first moon jalopy (a '71 softtop), the 77-pound lunar rover that tipped the scales at 462 pounds in Florida. The complexity and intricacy of components for manned spacecraft necessitate a reliability factor exceeding 99.9999 percent. That standard would permit a car on Earth to operate for a century without any malfunctions.

A Quiet Place for a Walk

On December 7, 1972, *Apollo 17*, the last of the program, became the first to be launched at night, lighting up the entire Cape. The mission was also the longest, at 301 hours, with a quarter of that time spent on the moon. Crewed by Gene Cernan, Ron Evans and scientist-astronaut Jack Schmitt, *17* was a thorough success, collecting 243 pounds of lunar samples. Still, there was a minor letdown along the way. At one point, Schmitt was foraging when he suddenly exclaimed, "Hey, there is orange soil! It's all over!" His obvious excitement, which was shared back home, stemmed from

NASA/Michael Light

the fresh appearance of the soil. Also, it was in a crater similar to volcanic vents on Earth. The suggestion was that there might have been volcanic activity on the moon as recently as 200,000 years ago, a far cry from the widely held view that Earth's satellite had been dormant for some 3 billion years. Tests revealed, however, that the soil was 3.71 billion years old, typical for a moon sample. Above: In this panoramic composite of photographs taken by Cernan with a Hasselblad, Schmitt is making his way around Split Rock. The mountain to the right rear, known as the South Massif, is about five miles away. The East Massif may be seen on the left.

NASA (2)

An Historic Program Draws to a Close

The Apollo missions cost $25 billion, surely a considerable sum but rather modest compared with the results. There were 11 manned journeys in which 12 men walked on the moon—one, the first ever to do so. Nearly a half ton of samples was brought back, and some things were left behind. At right, Charles Duke placed this snapshot of his family on the lunar surface before departing with *Apollo 16.* Above, Jack Schmitt admires the Stars and Stripes, with Earth beyond. Before he and Cernan—the last men to walk on the moon—finished their work for *Apollo 17,* they set eight explosive charges that would, along with the impact of the lunar module after being jettisoned, be registered later as part of a seismographic experiment. These final tests prompted one space historian to write, "The Apollo program was leaving the moon with nine bangs and no whimpers."

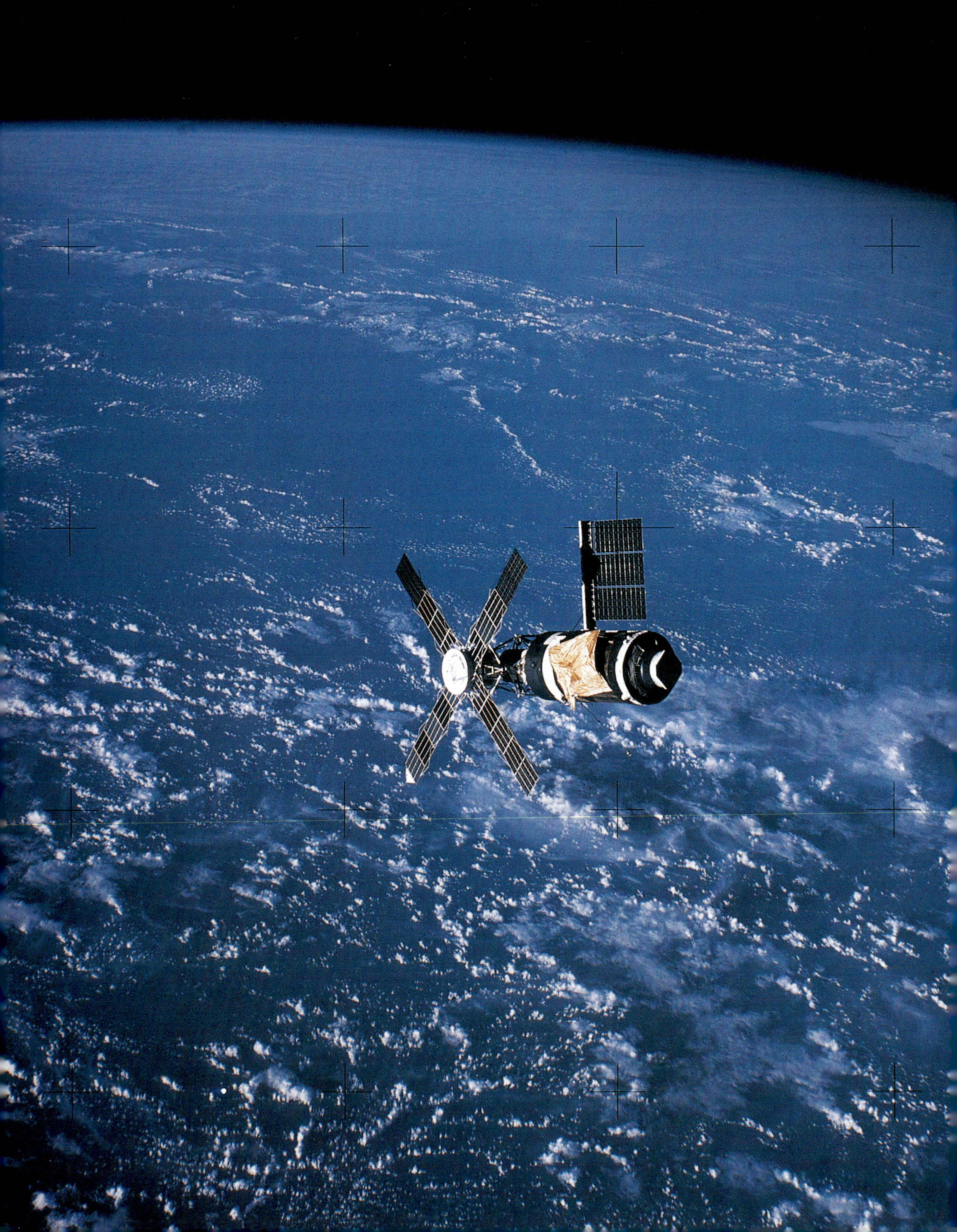

NASA (2)

Who Can Fix the Jammed Wing? The Handyman Can.

The first unmanned U.S. space station went into orbit on May 14, 1973. Originally designed to gather info for the Apollo program, it was called Skylab for its role as a laboratory. In its ascent, an external shield tore off, taking with it one solar-panel wing and jamming the other. Nine days later, the manned *Skylab 1* mission arrived on the scene. Things did not proceed smoothly. Attempts at docking failed eight times. Once the vessels were joined, Pete Conrad and Joe Kerwin (above, in zero gravity) embarked upon a perilous four-hour space walk to free the jammed wing. (Left: Golden foil was attached as a sunshade.) Their job was made harder whenever they moved into Earth's shadow and their only light was that spilling from Skylab's hatch. Finally, success. But of course: Conrad had long been known as Mr. Fixit.

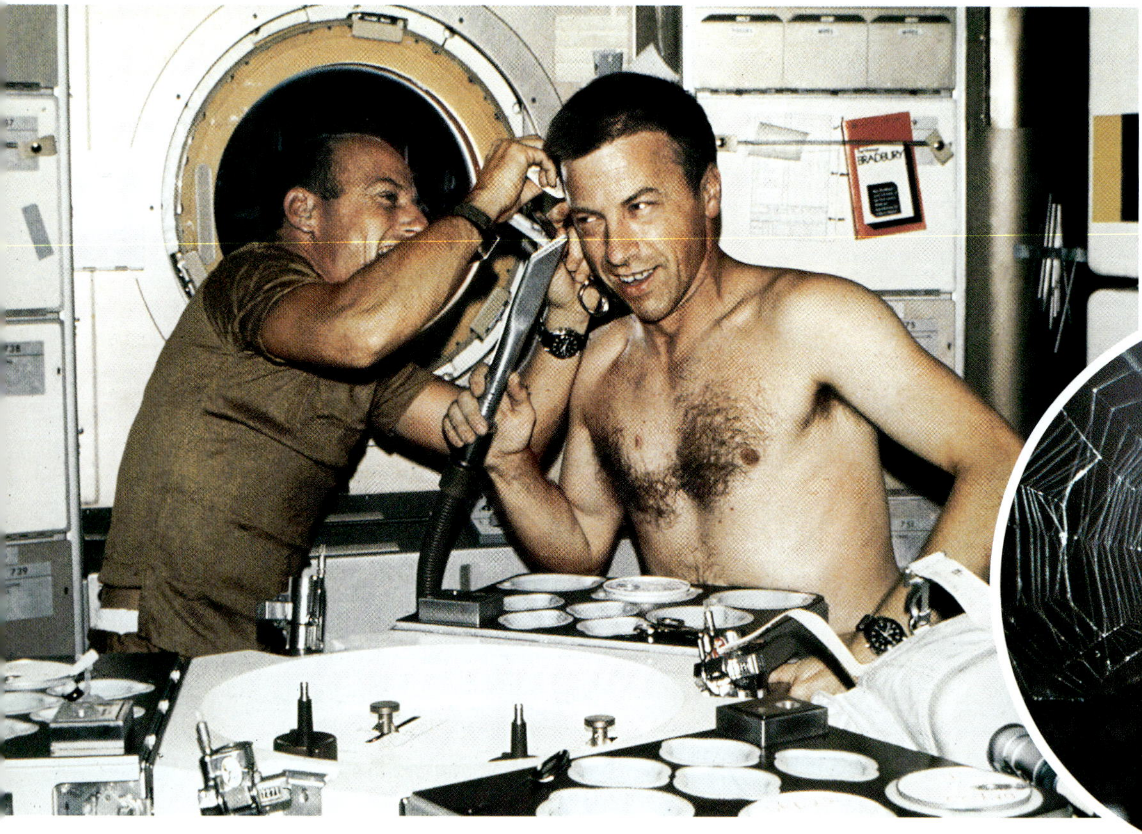

NASA (4)

Just Another Day at the Space Station

Skylab hosted experiments in geography, engineering, biomedicine and other disciplines. The physical and psychological effects of zero gravity and isolation on a human being were also of prime importance. The nine men who went up on the three Skylab missions were up to the challenge. Top left: Conrad takes a shower. Sounds simple, but owing to weightlessness, it takes an hour because he must vacuum water from his body and the curtain. Bottom left: Paul Weitz collects clippings as Conrad gives him a trim. Above: *Skylab 2*'s Jack Lousma tries out a jet-powered flier. Left: Arabella, another earthling aboard *Skylab 2,* in an experiment suggested by a Massachusetts high school girl, shows she can do her thing even in zero gravity (although she died of dehydration). Skylab was finally undone by increased sunspot activity, which messed with its formerly pristine orbit. The wayward space station burned up over the Indian Ocean in 1979.

Bettmann/Corbis

For the Love of Science and Adventure

These Soviets were the first men to crew a space station. Left to right: Georgi Dobrovolskiy, Viktor Patsayev and Vladislav Volkov were on *Soyuz 11,* which in 1971 docked with the orbiting Salyut. They remained for 22 days. While reentering Earth's atmosphere, the capsule depressurized, killing the cosmonauts. When President Nixon saw a photo of Dobrovolskiy's mourning 12-year-old daughter Maria, he wrote her: "I know your father would want you to face his loss with a strength and courage to match his own."

TASS/AP

NASA (4)

Miracle in Space

In a photograph taken on July 17, 1975, from the orbiting American spacecraft *Apollo,* a sun-glistened Soviet *Soyuz* glides majestically toward the first rendezvous and docking of vessels from different nations. Given that relations between the two superpowers were anything but cordial at the time, the event marked a diplomatic as well as scientific coup. As the ships drew within 50 miles of each other, Moscow Control sent this message in English: WE CROSS OUR FINGERS. Clockwise from top left, the two skippers, Tom Stafford and Alexei Leonov, meet in the linking hatchway; Leonov shows off his artistic side with a sketch he drew of Stafford. When the crews met in space, they exchanged bear hugs and gifts such as flags. Here, Stafford and Deke Slayton seem smitten with these tubes of Russian vodka.

Charles L. Trainor/Miami Daily News

The 1980s

The Shuttle Lifts Off

The theory was elegant, straightforward, even simple: A spaceship with wings would zip into the void and return, again and again, thereby making space as easily and cheaply accessible as, say, London or Paree. But after nearly $10 billion had been spent—a 30 percent cost overrun—and a ton of scorn heaped, many wondered if the idea would ever fly. The craft assigned to silence the naysayers was named *Columbia*.

Joe McNally

Bettmann/Corbis (2)

The all-important tiles that will protect the ship during reentry are applied to *Columbia*'s immense hide. The 40-mile commute to the Edwards Air Force Base in California takes nine hours. There, *Columbia* boards a Boeing 747 for a cross-country lift to Cape Canaveral, from which it will launch.

The goals of the American space program had been ticked off like so many items on a preliftoff checklist. Send a satellite up. Check. Send an ape up. Check. Send a man up. Check. Catch the Soviets, beat the Soviets to the moon, rocket past the Soviets. Check, check, double-check.

In its maturity, NASA needed to identify new ends that would be worth the means. The fact that the moon hadn't hosted an earthling since 1972, and that there were no plans to land again on the lunar surface in the coming decade, was telling. In his 1963 speech to the U.N., President Kennedy had invoked the English mountain climber George Mallory, who, when asked why he was going to Everest, said, "Because it is there." Kennedy said, "Well, space is there, and we're going to climb it. And the Moon and the planets are there and . . ." With the cold war all but ended and the U.S. economy no longer maintaining its reliable post–World War II pace, such bold but venturesome rhetoric no longer carried the day. Justification was needed for the next generation of space missions, particularly when technological advances had reached a point where a computer chip, or perhaps one of those 1950s monkeys, could perform many aboard-ship duties while Houston pulled the strings.

President Richard M. Nixon's Office of Management and Budget was hovering over NASA like a vulture in the early 1970s, and

was at one point considering the cancellation of *Apollos 16* and *17,* as well as a halt to development of a reusable space shuttle system, something that had been ongoing since 1969. It may well be that the motivating factor behind Nixon's January 1972 approval of funds for the shuttle project was to boost sagging employment figures in states that were considered crucial in that year's presidential election. Whatever the case, the shuttle was green-lighted, and plans for a fleet of orbiters were drawn. The first airplane-like space vehicle to be launched would be *Columbia,* in April 1981.

Although the shuttle did not prove to be the easy, inexpensive route to routine space exploration that had been hoped—some planners had envisioned a turnaround of days, not months or years—it rekindled interest in NASA's sacred grail: a permanent orbiting station from which to conduct research and leap into deeper space. In 1984, sounding positively Kennedy-esque, President Ronald Reagan said, "America has always been greatest when we dared to be great. We can reach for greatness again. We can follow our dreams to distant stars, living and working in space for peaceful, economic and scientific gain. Tonight I am directing NASA to develop a permanently manned space station and to do it within a decade."

Lofty goals were again in sight. But beyond the horizon lay unexpected tragedy, and all debates were far from over.

Sam Walton

NASA

Dark Skies or Fair?

The inaugural 1981 mission of *Columbia,* with space veterans John Young and Bob Crippen as the only two aboard, was judged a huge success. Gone overnight were the "jokes about America's 'space lemon' and 'flying brickyard,'" *Time* magazine reported in a cover story. "So long delayed, so widely criticized, *Columbia*'s flight should finally put to rest any doubts that there will one day be regular commuter runs into the cosmos." As the shuttle era progressed, it became clear that the flawless first flight had done no such thing. Workhorse members of the early shuttle lineup, including *Challenger* (left, as an hour-long electrical storm delays the launch of its third flight in 1983) and *Columbia* (preparing to land in the California desert after its second mission in '81), generally fared well but faced the stark reality that, in such a complex system traveling at such remarkable speeds, a success rate of 99 percent would be admirable. In other words, one day, something was bound to happen.

Sovfoto

More Barriers Falling

Even in the 1980s, with manned space exploration in its third decade, there were firsts registered by astronauts and cosmonauts. Above: During a Soviet mission in 1982, Svetlana Savitskaya becomes the second woman in space and the first female member of a coed crew. Two years later, this daring daughter of a World War II flying ace (Svetlana had made 450 parachute jumps before she was 17) became the first woman to walk in space. In back-to-back U.S. shuttle missions in 1983, Sally Ride (opposite) and Guy Bluford Jr. became the first American woman and the first African American in space. Bluford would go back into space three times, logging 688 hours in all. Astrophysicist Ride would return to academia, while retaining her status as a role model for all little girls. She has written four children's books about space.

NASA (2)

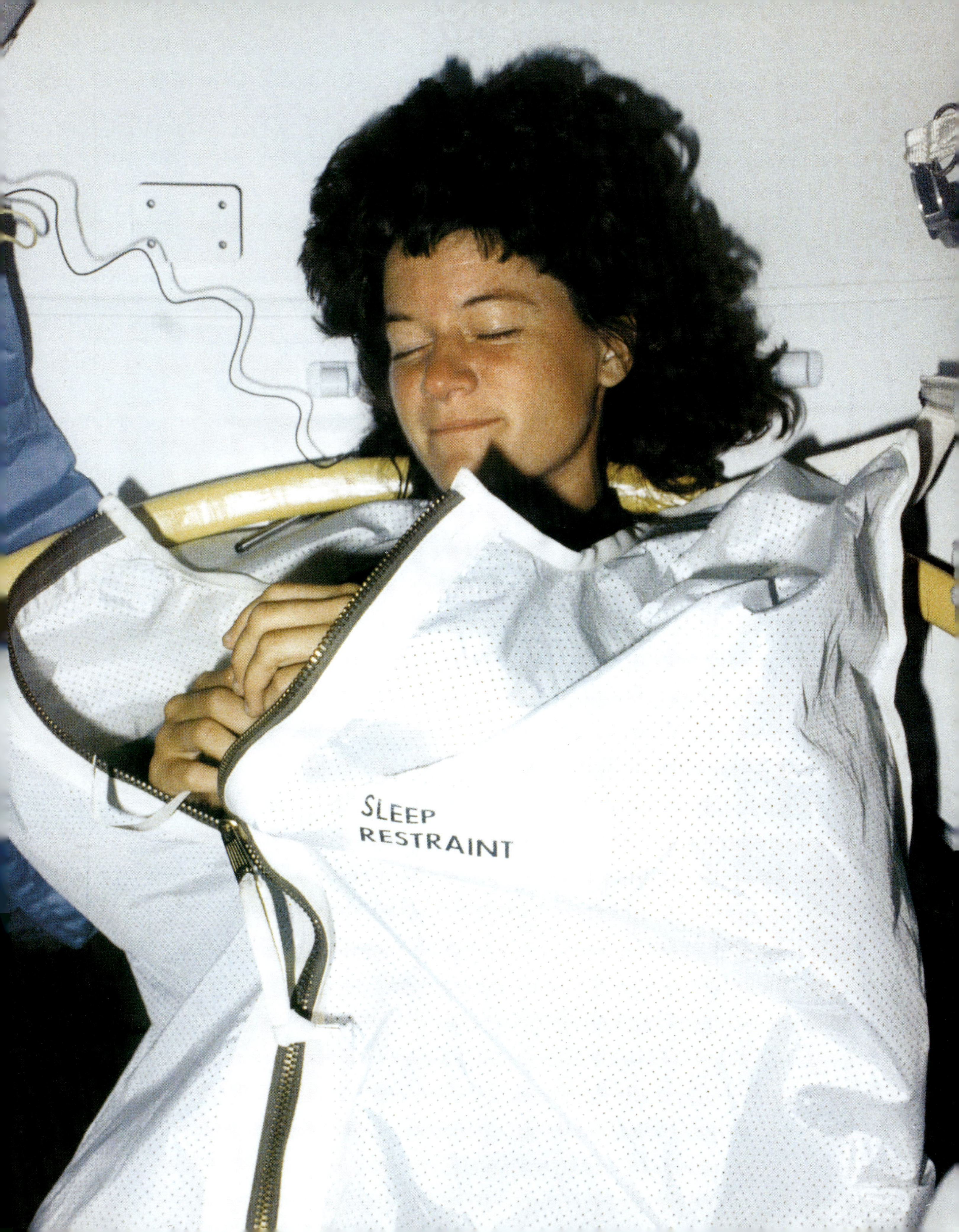
SLEEP
RESTRAINT

Bruce McCandless/NASA

NASA (2)

McCandless Unbound

Of the shuttle program's first fleet of orbiters, one seemed bent on making history. *Challenger* was the vehicle for the groundbreaking voyages of Ride and Bluford and, in early 1984, that of Bruce McCandless, who became the first free-flying human in space. At 8:25 a.m. on February 7, McCandless, an astronaut since 1966 but a space rookie, stepped—untethered—from the shuttle into the void. Using a nitrogen-propelled backpack, he jetted 300 feet from *Challenger* at a rate of one foot per second, even as he and the spaceship orbited Earth at about 17,000 mph. "Everyone thinks that space is black, but I know from firsthand experience that it's *really* Navy blue," said the 46-year-old naval captain. Above: McCandless's photo of his colleague Bob Stewart, working in *Challenger*'s payload bay.

NASA (2)

Tobey Sanford

A New Kind of Space Traveler

The shuttle, in NASA's own words, "changed the sociology of space flight." Large crews now featured traditional spacemen and also "payload specialists—who are not even necessarily career astronauts." Early in the next millennium, this definition would devolve to include—in the Russian program, at least—wealthy tourists and teen-dream pop stars. But in the shuttle's first years, it usually meant scientists conducting microgravity experiments. In the mid-1980s, NASA felt that another step in demystifying space travel could be taken, and asked America's teachers if one of them might want to ride the shuttle. Christa McAuliffe, a 37-year-old high school social studies teacher from Concord, N.H., was selected above 11,145 other applicants, and went to Houston for training in late 1985. Above, on a NASA KC-135 turbojet diving at 35,000 feet, McAuliffe experiences the feeling of weightlessness. Left, Commander Dick Scobee leads the seven-member crew; McAuliffe is third from the rear.

The Worst Imaginable Scenario

"I collect newspapers and magazines," McAuliffe said in an interview during the run-up to her planned weeklong mission. "In my attic I have the LIFE magazine of the men walking on the moon, and I have a box of newspapers with articles about my own flight. When I'm 60, maybe, I'll look at my pile of papers and wonder, 'What really happened that year?'" Scobee, too, felt theirs was to be a flight for posterity, as he told McAuliffe, "Missions keep going up every month, but the teacher in space is unique. *You're* the reason we'll be remembered." While he was right that McAuliffe's participation greatly raised the profile of the 10th flight of the orbiter *Challenger*—prompting millions to tune in to live TV coverage of a rocket launch as had earlier generations of Americans—he was tragically wrong in his prediction: The *Challenger* will always be remembered because, 73 seconds after liftoff on January 28, 1986, there was an horrific explosion in the sky. Below: Moments earlier, Christa McAuliffe's family had heard a P.A. warning of a *Challenger* malfunction. Here, the unbearable nightmare is playing out before their eyes.

Keith Myers/The New York Times

Michelle McDonald

Diana Walker

Enrico Ferorelli

In the Aftermath, Tears and a Search for Answers

Searching for debris and the victims' bodies took time, and memorials and funerals were spaced over months. Above, June Smith, the widow of Navy Capt. Mike Smith, is consoled by her daughter during the Houston service. Below, in April, prior to transfer to locations across the country, a procession of the crewmembers' coffins at Dover Air Force Base in Delaware. The shuttle's wreckage was sifted through in Florida as part of an investigation by a presidential commission, which placed blame on a design flaw in the solid-rocket boosters' O-ring seal. Smith, apparently, had seen flames leap past his window; his final utterance as heard by Mission Control was, "Uh-oh!" If any of the crew were alive during the two-minute, 20-second plunge to the ocean, they were killed upon the 207-mph impact.

Mark Meyer

Unmanned Explorers

The *Challenger* disaster forced an immediate grounding of all shuttles, and no American would return to space for more than two years. But NASA's program didn't shut down. Probes, which had been sending back images since the '60s, probed on—transmitting ever more amazing pictures from ever farther out. The mission: to go where no man had gone before.

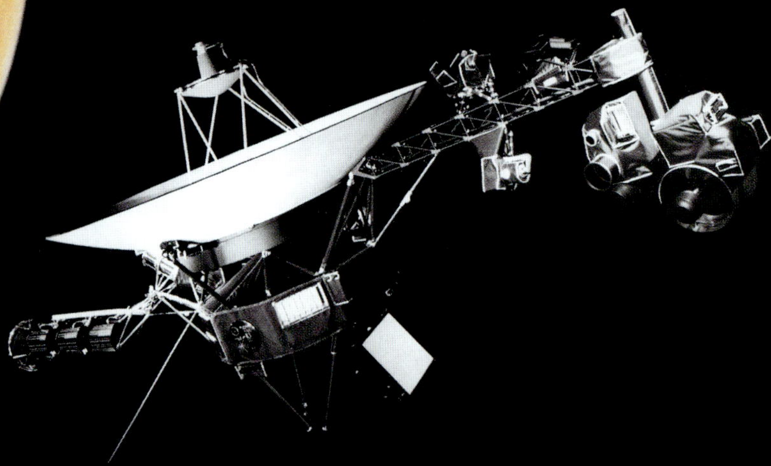

Venus Is Really Hot, and Mars Is Truly Cold

Probes, like the Voyager at left, relayed not only terrific eye candy but also information that allowed scientists to confirm or refute theories about the solar system. For instance, regarding our near neighbors: Temperatures on Venus (right and above, in images based on Magellan's '91 data) are now gauged at more than 800°F, and the rocky Martian surface (viewed by Viking 2 in '76, at left, and probed by Pathfinder's rover in '97, below) can dip below –200°F.

Voyager spacecraft: J.R. Eyerman; JPL/NASA (4)

In Our Solar System's Far Reaches, the Strange and Wondrous Realm of the Gas Giants

Probe data confirmed a prevailing theory of how the solar system formed: A nebula of dust and gas rotating around the sun condensed into individual bodies. The gaseous covers of the inner planets were ripped from their solid cores by a wind of particles from the sun, while the thick atmospheres of the outer planets were protected from the solar storm by size and distance. Jupiter (above, with moons Io and Europa, seen from Voyager 1, 1979), Neptune (left, Voyager 2, 1989) and Saturn (top right, Cassini, 2002, and bottom, Voyager 1, 1980) remain, like the sun itself, balls of volatile gases.

JPL/NASA (4)

Many Moons of Many Kinds, Each with a Mysterious Tale to Tell

Voyager's imaging team saw craters on Jupiter's moon Ganymede (above, in 1979) and knew meteors had once swept through the solar system. This supported the theory that all planets and moons formed at the same time, 4.6 billion years ago. Opposite, a Voyager 2 portfolio: Neptune's Triton (1989) and, from top, Saturn's Enceladus and Hyperion ('81) and Uranus's Ariel ('86). Below: A volcano on the Jovian moon Io blows 100 miles high in 1979.

JPL/NASA (6)

The 1990s

New Horizons

Certain events hearkened back to the space race of yore; others seemed as far-out as tomorrow. Boys paddling off Florida's coast gaze in awe—as boys had 36 years previously—when none other than John Glenn returns to Earth's orbit. Meanwhile, robots roamed the Martian surface and a fantastic telescope probed deep, deep space. The question was inescapable: To whom does the future belong, man or machine?

Gregg Newton/Reuters/Getty

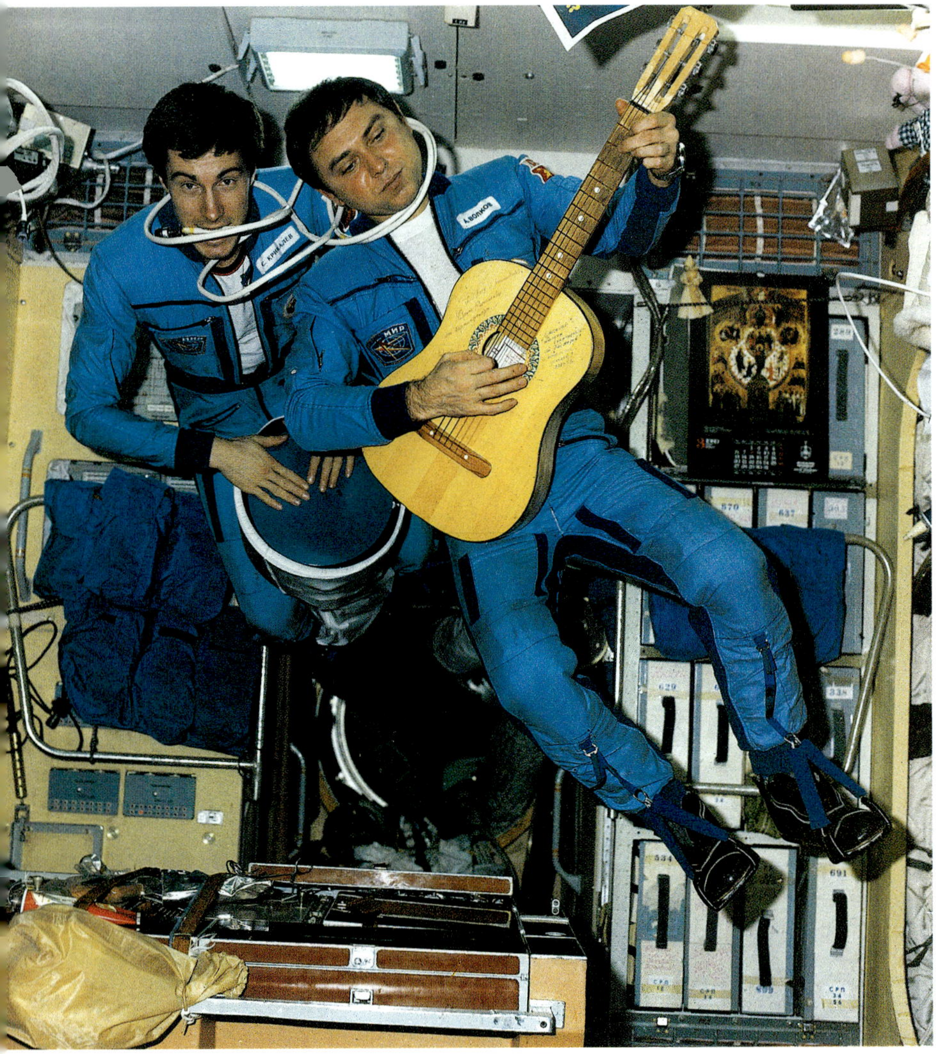

TASS/Sovfoto

Suddenly, the whole world's harmonizing. Sergei Krikalev, on percussion, and Alexandr Volkov spend time on Mir as Soviets in 1988 and '89, meet again on Mir as Soviets in 1991, then return to Earth as Russians in '92 (the U.S.S.R. vanishes while they're away). In 1997, Mike Foale (near right), a British-born Yank, is in Star City, Russia, training with his new best cosmonaut pals for a stint on Mir, at the time the world's only orbiting space station.

Only weeks before the 1990s rolled in, a physical manifestation of the cold war that had stood for three decades was at last breached when the gates to the Berlin Wall were finally opened. While the human drama of sudden freedom was the stuff of headlines everywhere, the gripping theater of humans in space—Eagle versus Bear—was reconfiguring as a faceless plexus of payload specialists, exotic materials and labyrinthine machinery, and drifting ever further away from the days of the glamorous test pilot. A year after the Wall was dismantled—1990—the Soviet Union, too, was disassembled. Its once-mighty space program was left to Russia alone, which meant that operations would perforce be on a considerably reduced scale. Japan and the European Space Agency, however, joined in helping pick up slack.

In the spirit of the above victories for democracy, all now was quiet on the international space front—fraternal rather than confrontational—but this very fraternity, in lieu of sworn enmity, seemed to reduce any obsession with manned events, ones that might lead to further extraterrestrial "conquests." Some asked: Should man even be in space?

Although there would be no new footprints on the moon, lunar interest chugged along with flybys and surveying. Increasingly, Mother Earth herself became a prime object of attention, through Spacelab, Mir and other missions. And after growing pains, the wunderkind of telescopes, Hubble, brought into focus places we might hope to visit in time to come. As if that weren't enough, Pope John Paul II acknowledged that the Vatican had erred so many years ago when it condemned Galileo for supporting the Copernican Theory, viz., that the earth revolves around the sun.

So, earthlings still did long to visit space, even if our need was perhaps becoming more clinical than passionate.

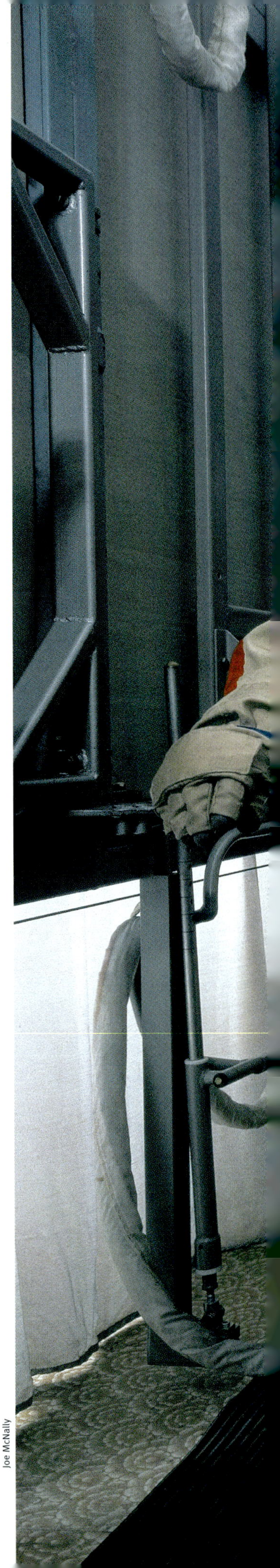

Joe McNally

NASA

No Time for Joking

The Soviets' Mir station was sent to space in 1986, and the old, abandoned Mir was sent to the Pacific Ocean in a fiery blaze in 2001. In the intervening 15 years, Mir performed yeoman service, hosting scores of spacefolk from more than a dozen countries. But by the late 1990s, Mir was tired, and the shaky station was fodder for talk-show comics. In 1997 a fire was followed by a collision with a resupply ship. During the latter crisis, loss of air pressure forces Mir's crew to save power; on Earth, flight director Vladimir Solovyov frets and frogmen act out repairs on a Mir replica. With a crew in trouble, the laughter ceased.

Sergei Karpukhin/AP

Dima Korotayev/Reuters

Reuters

Bandages Applied, Mir Limps Away

Mike Foale and Russians Alexandr Lazutkin and Vasily Tsibliyev (below, from left) were forced to abandon experiments, listen to instructions from Mission Control and do repair work. They secured the station, and in August 1997, Tsibliyev and Lazutkin (above) are relieved to be back on terra firma after exiting the *Soyuz*. In October, *Atlantis* returns Foale to his daughter, Jenna, and son, Ian. America's involvement with Mir ended in 1998, and the last Russians left the station in June 2000. That October, the grand new International Space Station welcomed its first crew.

AP

George Shelton/NASA

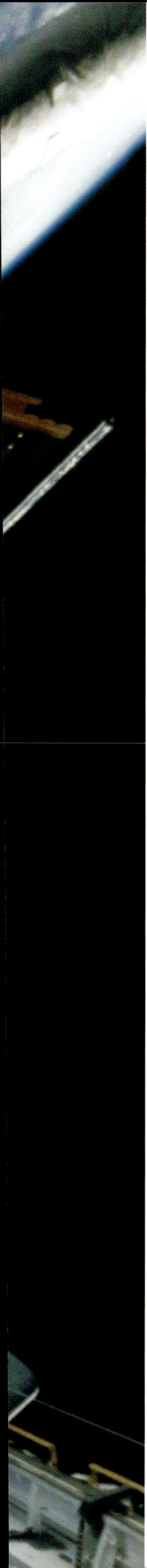

A Machine in Need of a Man

The shuttle has been the Hubble Space Telescope's best friend. In April 1990, this extraordinary optical device was taken aloft by *Discovery* and set into Earth's orbit, from whence it would peer into deep space and send back images and data providing clues to the origins and development of the universe. That was the plan. But it was soon determined that the Hubble was impaired. In 1993, the crew of *Endeavour* fixed the scope during a record five EVA sorties. Again in February 1997, the Hubble required servicing, and this time it is Steve Smith (left), working outside *Discovery* (that's Australia in the background), who helps repair the machine and redeploy it. In 1999, Smith was back on the job for yet a third Hubble spruce-up. The yield of all this human toil has been everything NASA dared hope for—and then some. Year after year, the Hubble has made the world a smaller place. Right: The first true-color photo of Jupiter, taken in 1991. Below: Part of the Eagle Nebula, a star-forming region 7,000 light-years away, 1995.

Jeff Hester & Paul Scowen/NASA

NASA (2)

. . . And to Space Ye Shall Return

Given that our society has been radically altered by scientific advances permitting people to live longer and longer, it seemed entirely appropriate when John Herschel Glenn Jr., at the age of 77, volunteered in 1998 to return to his old stomping grounds: the really wild blue yonder. As a veteran of 149 combat missions in World War II and Korea, as a pilot who set a transcontinental speed record in 1957, as a flier with more than a year spent in the air, as the man in the first American orbital mission, this U.S. senator and retired Marine colonel was one septuagenarian, if there was ever going to be one, who could handle outer space. Fine, said NASA's chief administrator Dan Goldin. "He deserved a second flight. But I told him two things: The project would have to be real, peer-reviewed science, and he'd get no freebies on the astronaut physical." Only a fool would've bet against this man of steel. Right: The crew of space shuttle mission *STS-95* flanks the grand old man.

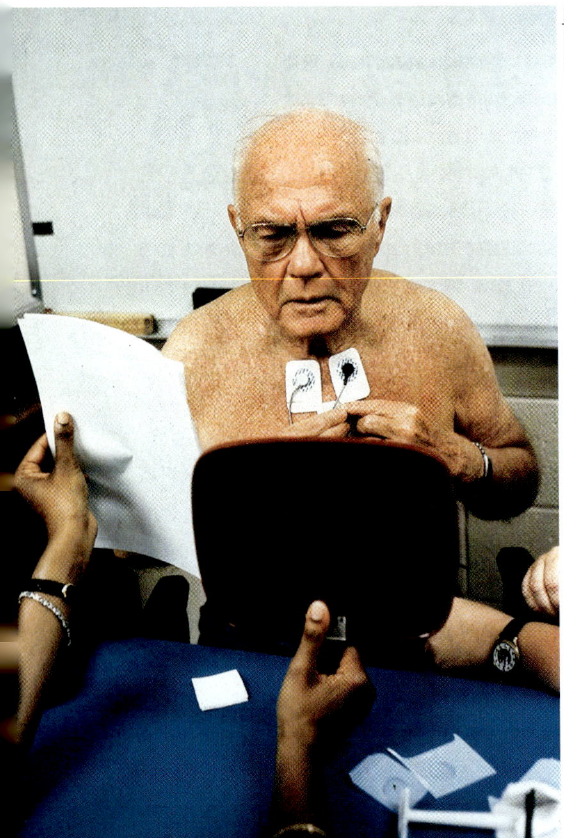

Ralph Morse

Joe McNally/National Geographic for NASA

Joe McNally/National Geographic for NASA (3)

Robert Markowitz/NASA

A Jewel in the Senior Citizen Crown

In the 1960s, NASA had noticed that certain symptoms of space flight echoed those noted by the elderly. Glenn's job on the space shuttle *Discovery* was to study similarities between space travel and aging. Clockwise from left: helping payload specialist Chiaki Mukai (it was a multinational crew) with her suit; practicing survival methods; performing an egress exercise that was "actually kinda fun"; a wave to the press; carrying out an experiment in the Spacehab facility aboard the shuttle. *Discovery* landed on November 7, after a nine-day flight, and Mr. Spaceman set off for yet another of his ticker-tape parades. Said Glenn: "A lot of attention focused on me, but in the end I really was just payload specialist number two."

NASA

"The Columbia Is Lost"

A second tragedy in the space shuttle program had a galvanizing effect on the American public, reminding many why it is that we go into space.

Gene Blevins/Daily News

Dr. Scott Lieberman/Tyler Morning Telegraph/AP

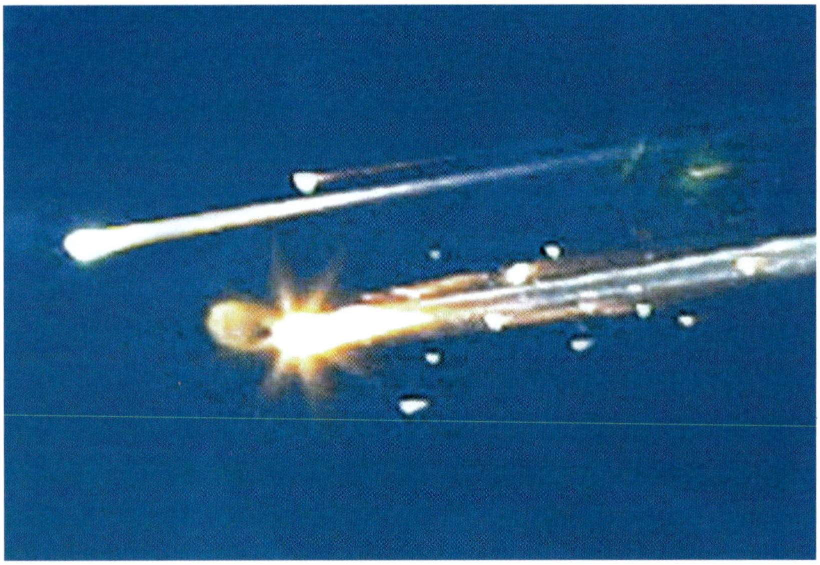

February 1, 2003. A Saturday morning. Many of us were barely aware it was up there—barely aware, or unaware. Others knew, surely, but didn't think much about it. The Challenger *disaster was so many years ago, and the string of shuttle successes had again grown so long—15 years of smooth sailing, 87 missions since that shocking explosion—and we Americans had again grown complacent, if not blasé, about the shuttle. Perhaps, even, about space travel altogether.*

Besides, we would be at war soon. Bin Laden was out there, somewhere. The economy was in the dumps. There were storms heading in and . . . and then word spread house to house on that Saturday morning. "Turn on your TV . . . Turn on your radio . . . Something's happened with the shuttle . . . Something's happened with the shuttle again."

Again. That had to mean another shuttle orbiter had crashed or blown up. Why else would we be asked to turn on the set to see what had happened?

Which one was it? There are four, aren't there?

Above: Six minutes before its fateful end, the shuttle streaks over the Owens Valley Radio Observatory, north of Bishop, Calif. Opposite: Debris is visible as *Columbia* disintegrates high in the morning sky over Tyler, Tex.

Columbia, Atlantis . . . Discovery, *is it? . . . and . . .*

What had we all heard about the shuttle recently? Something, yes? There was a rich American fellow—what was his name?—who had bought himself a ride into space. Was he on a shuttle? And that singer from 'NSync was trying to hop the shuttle, wasn't he? Goodness . . . Was he up there?

No, he was not. NASA, to its credit, has firmly stated that shuttle rides are not for sale. It was the thoroughly strapped Russian space program that in 2001 made Los Angeles–based financier Dennis Tito the first space tourist by selling him a trip to the International Space Station in a *Soyuz* capsule for $20 million. The Russians were only too happy to accommodate pop star Lance Bass, too, until his sponsors were late in paying for his ticket; at that point, they booted Bass out of Star City.

That the questions were asked, however—and that Tito and Bass were two of the more prominent space personalities of recent years—illustrates how far space exploration had fallen in the general consciousness. The shuttle might have been up there

NASA

or not at any given time. It might have been carrying scientists, some astronauts, maybe a politician or a teacher. Who knew? No news surely must be good news.

The bad news stunned us awake to the eternal realities of space travel. We recalled how perilous it was—how exciting, but also how frightfully dangerous. And when we learned more about the seven astronauts aboard *Columbia,* we remembered how brave, intrepid, dedicated, even heroic, were those men and women who dared venture into space.

It was, as most shuttle crews have been for years, an ethnically mixed team of men and women. There were both Americans and foreign guests: a microcosm of the world. Mission commander Rick Husband, 45, got his pilot's license while still a teenager in Texas. He used to put on his astronaut suit at Halloween and pose with kids at the church where he worshiped, outside of Houston. San Diego–born Willie McCool, 41, *Columbia*'s pilot, had two master's degrees, four Navy medals and almost 3,000 flight hours when he joined NASA in 1996. Husband left behind a wife and two children, McCool a wife and three kids.

Ilan Ramon, 48, the payload specialist, was a married father of four and a colonel in the Israeli Air Force. In 1981, at age 26, he had piloted one of eight warplanes into Iraqi territory on a successful mission to destroy a nuclear reactor. Another *Columbia* crewmember from abroad was mission specialist Kalpana Chawla, 41, from India. In her homeland, which doesn't nurture achievement by women, this shuttle veteran—she first flew in 1997—was a hero to girls.

Mission specialist Dave Brown, 46, was a physician, flight surgeon and naval aviator before he became an astronaut. Mission specialist Laurel Clark, 41, was a submarine medical officer and a flight surgeon before she became an astronaut. Payload commander Michael Anderson, 43, was an Air Force lieutenant colonel who was one of only seven black astronauts out of NASA's 80. These were

Top left: Ramon gazes outward from *Columbia.* In Wyoming in August 2001, the crew trains, and bonds (from left): McCool, Brown, Ramon, Husband, Clark, Anderson and Chawla. Barbara Anderson mourns her son in Houston, while John and Annie Glenn weep for the crew at a memorial service in Washington, D.C.

John Kanergieter/NOLS

Roberto Schmidt/AFP

ing impact. It made many, if not all, Americans long for space once more, and for what going into space represented.

"We are that part of creation which seeks to understand all creation. We find the best among us, send them forth into unmapped darkness, and pray they will return," said President George W. Bush at a memorial for the *Columbia*'s crew. "This cause of exploration and discovery is not an option we choose. It is a desire written in the human heart."

people of accomplishment as well as courage, sterling representatives of all the accomplished, courageous space travelers that had preceded them. Rick Husband saluted some of those bygone astronauts when *Columbia* observed the January anniversary of the 1986 *Challenger* tragedy while in space: "They made the ultimate sacrifice, giving their lives to their country and mankind. Their dedication was an inspiration to each of us."

The lives and deaths of those aboard *Columbia* had a similarly inspiring effect on the American people. We remarked at the dedication of the astronauts, what extraordinary things they had been doing without receiving the proper attention. This awful situation—the disintegration of the first space shuttle ever to fly—had a remarkable and surpris-

Pablo Martinez Monsivais/AP

The Future?

NASA/IMAX

Three Men Still Aloft

As had happened in 1986 in the aftermath of *Challenger*, the *Columbia* tragedy prompted a reevaluation of America's mission in space. Everyone from pilots to politicians started arguing for or against manned flight. What escaped many was that, above the din, manned flight continued. In 1998, assembly began in outer space on a phenomenal cooperative venture, the International Space Station. Five years later, U.S. Navy Capt. Ken Bowersox, flight engineer Don Pettit and the Russian flight engineer Nikolai Budarin were orbiting Earth in the still-to-be-finished station when *Columbia* disintegrated. The questions that circled with them: After they head home in May, aboard a Russian vehicle, how long will it be before a shuttle returns to the station? Will manned ships—American? Russian? Chinese?—ever again set sail for the moon or some distant planet? Or will humans be earthbound, yielding space to robots? Whither man in space?